Fourier Transforms:
Principles and Applications

Fourier Transforms: Principles and Applications

Editor: May Hooper

NY RESEARCH
P R E S S

New York

Published by NY Research Press
118-35 Queens Blvd., Suite 400,
Forest Hills, NY 11375, USA
www.nyresearchpress.com

Fourier Transforms: Principles and Applications
Edited by May Hooper

International Standard Book Number: 978-1-63238-678-6 (Hardback)

Cataloging-in-Publication Data

Fourier transforms : principles and applications / edited by May Hooper.
    p. cm.
Includes bibliographical references and index.
ISBN 978-1-63238-678-6
1. Fourier transformations. 2. Fourier analysis. I. Hooper, May.
QC20.7.F67 F68 2019
515.723--dc23

Contents

Preface

This book was inspired by the evolution of our times; to answer the curiosity of inquisitive minds. Many developments have occurred across the globe in the recent past which has transformed the progress in the field.

The Fourier transform is an important area of mathematics. It is an extension of the Fourier series, in which all periodic functions are expressed as the sum of sine and cosine wave functions. The Fourier transform is relevant when the time period of a represented function approaches infinity. This mathematical tool decomposes a signal, which is a function of time, into its constituent frequencies. The Fourier transform obtained from this decomposition is a complex-valued function of frequency. To recover the original function of time, an inverse Fourier transformation can be operated on the decomposed function. The basic properties of the Fourier transform are linearity, invertibility and periodicity, etc. Fourier transforms have applications in solving differential equations, in quantum mechanics and signal processing. Besides these, it is of great significance in spectroscopy, especially nuclear magnetic resonance, magnetic resonance imaging and mass spectroscopy. This book provides comprehensive insights into the principles and applications of Fourier transforms. Different approaches, evaluations, methodologies and advanced studies on Fourier transforms have been included in this book. Coherent flow of topics, student-friendly language and extensive use of examples make this book an invaluable source of knowledge.

This book was developed from a mere concept to drafts to chapters and finally compiled together as a complete text to benefit the readers across all nations. To ensure the quality of the content we instilled two significant steps in our procedure. The first was to appoint an editorial team that would verify the data and statistics provided in the book and also select the most appropriate and valuable contributions from the plentiful contributions we received from authors worldwide. The next step was to appoint an expert of the topic as the Editor-in-Chief, who would head the project and finally make the necessary amendments and modifications to make the text reader-friendly. I was then commissioned to examine all the material to present the topics in the most comprehensible and productive format.

I would like to take this opportunity to thank all the contributing authors who were supportive enough to contribute their time and knowledge to this project. I also wish to convey my regards to my family who have been extremely supportive during the entire project.

Editor

1

Henstock-Kurzweil Integral Transforms and the Riemann-Lebesgue Lemma

Francisco J. Mendoza-Torres, Ma. Guadalupe Morales-Macías,

Salvador Sánchez-Perales and Juan Alberto Escamilla-Reyna

Additional information is available at the end of the chapter

1. Introduction

Let f be a function defined on a closed interval $[a, b]$ in the extended real line $\overline{\mathbb{R}}$, its Fourier transform at $s \in \mathbb{R}$ is defined as

$$\hat{f}(s) = \int_a^b e^{-ixs} f(x) dx. \tag{1}$$

The classical Riemann-Lebesgue Lemma states that

$$\lim_{|s| \to \infty} \int_a^b e^{-ixs} f(x) dx = 0, \tag{2}$$

whenever $f \in L^1([a, b])$.

We consider important to study analogous results about this lemma due to the following reasons:

- The classical Riemann-Lebesgue Lemma is an important tool used when proving several results related with convergence of Fourier Series and Fourier transform. In turn, these theorems have applications in the Harmonic Analysis which has many applications in the physics, biology, engineering and others sciences. For example, it is directly applied to the study of periodic perturbations of a class of resonant problems.

- An important problem is to consider an orthogonal basis, different to the trigonometric basis, and study the Fourier expansion of a function with respect to this basis. In this case, it is obtained an expression in this way

$$\int_a^b h(xs)f(x)dx. \tag{3}$$

- In some cases the expression (1) exists and the expression (2) is true even if the function f is not Lebesgue integrable.

Thus, a variant of the Riemann-Lebesgue Lemma is to get conditions for the functions f and h which ensure that (3) is well defined and satisfies

$$\lim_{|s|\to\infty} \int_a^b h(xs)f(x)dx = 0. \tag{4}$$

Some results of this type and related results are found in [1], [2], [3], and [4].

In the space of Henstock-Kurzweil integrable functions over $\mathbb{R}$, $HK(\mathbb{R})$, the Fourier transform does not always exist. In [5] was proven that $e^{-i(\cdot)s}f$ is Henstock-Kurzweil integrable under certain conditions and that, in general, does not satisfy the Riemann-Lebesgue Lemma. Subsequently, it was shown in [6], [3] and [4] that the Fourier transform exists and the equation (2) is true when $-\infty = a$, $b = \infty$ and f belongs to $BV_0(\mathbb{R})$, the space of bounded variation functions that vanish at infinity. A special case arises when f is in the intersection of functions of bounded variation and Henstock-Kurzweil integrable functions.

There exist Henstock-Kurzweil (HK) integrable functions f which $f \in HK(I) \setminus L^1(I)$ such that (2) is not fulfilled, when I is a bounded interval. In [7], Zygmund exhibited Henstock-Kurzweil integrable functions such that their Fourier coefficients do not tend to zero. In [8] are given necessary and sufficient conditions in order to $\int_a^b f(x)g_n(x)dx \longrightarrow \int_a^b f(x)g(x)dx$, for all $f \in HK([a,b])$. Thus, we will prove that the Fourier transform has the asymptotic behavior:

$$\hat{f}(s) = o(s), \text{ as } |s| \to \infty,$$

where $f \in HK(I) \setminus L^1(I)$

Moreover in [9], Titchmarsh proved that it is the best possible approximation for functions with improper Riemann integral.

This chapter is divided into 5 sections; we present the main results we have obtained in recent years: [3], [4], [6] and [10]. In this section we introduce basic concepts and important theorems about the Henstock-Kurzweil integral and bounded variation functions. In the second part of this study we prove some generalizations about the convergence of integrals of products in the completion of the space $HK([a,b])$, $\widehat{HK([a,b])}$, where $[a,b]$ can be a bounded or unbounded interval. As a consequence, some results related to the Riemann-Lebesgue Lemma in the context of the Henstock-Kurzweil integral are proved over bounded intervals. Besides, for elements in the completion of the space of Henstock-Kurzweil integrable functions, we get a similar result to the Riemann-Lebesgue property for the Dirichlet Kernel, as well as the asymptotic behavior of the n-th partial sum of Fourier series.

In the third section, we consider a complex function g defined on certain subset of $\mathbb{R}^2$. Many functions on functional analysis are integrals of the form $\Gamma(s) = \int_{-\infty}^{\infty} f(t)g(t,s)dt$. We study the function Γ when f belongs to $BV_0(\mathbb{R})$ and $g(t, \cdot)$ is continuous for all t. The integral we use is Henstock-Kurzweil integral. There are well known results about existence, continuity and differentiability of Γ, considering the Lebesgue theory. In the HK integral context there are results about this too, for example, Theorems 12.12 and 12.13 from [11]. But they all need the stronger condition that the function $f(t)g(t,s)$ is bounded by a HK integrable function. We give more conditions for existence, continuity and differentiability of Γ. Finally we give some applications such as some properties about the convolution of the Fourier and Laplace transforms.

In section 4, we exhibit a family of functions in $HK(\mathbb{R})$ included in $BV_0(\mathbb{R}) \setminus L^1(\mathbb{R})$. At the last section we get a version of Riemann-Lebesgue Lemma for bounded variation functions that vanish at infinity. With this result we get properties for the Fourier transform of functions in $BV_0(\mathbb{R})$: it is well defined, is continuous on $\mathbb{R} - \{0\}$, and vanishes at $\pm\infty$, as classical results. Moreover, we obtain a result on pointwise inversion of the Fourier transform.

1.1. Basic concepts and nomenclature

We will refer to a finite or infinite interval if its Lebesgue measure is finite or infinite. Let $I \subset \mathbb{R}$ be a closed interval, finite or infinite. A *partition* P of I is a increasing finite collection of points $\{t_0, t_1, ..., t_n\} \subset I$ such that if I is a compact interval $[a, b]$, then $t_0 = a$ and $t_n = b$; if $I = [a, \infty)$, $t_0 = a$; and if $I = (-\infty, b]$ then $t_n = b$.

Let us consider $I \subset \mathbb{R}$ as a closed interval finite. A *tagged partition* of I is a set of ordered pairs $\{[t_{i-1}, ti], s_i\}_{i=1}^n$ where it is assigned a point $s_i \in [t_{i-1}, t_i]$, which is called a *tag* of $[t_{i-1}, t_i]$. With this concept we define the Henstock-Kurzweil integral on finite intervals in $\mathbb{R}$.

Definition 1. *The function $f : [a, b] \to \mathbb{R}$ is Henstock-Kurzweil integrable if there exists $H \in \mathbb{R}$ which satisfies the following: for each $\varepsilon > 0$ exists a function $\gamma_\varepsilon : [a, b] \to (0, \infty)$ such that if $P = \{([t_{i-1}, t_i], s_i)\}_{i=1}^n$ is a tagged partition such that*

$$[t_{i-1}, t_i] \subset [s_i - \gamma_\varepsilon(s_i), s_i + \gamma_\varepsilon(s_i)] \quad for\ i = 1, 2, ..., n., \tag{5}$$

then

$$|\Sigma_{i=1}^n f(s_i)(t_i - t_{i-1}) - H| < \varepsilon.$$

H is the integral of f over $[a, b]$ and it is denoted as

$$H = \int_a^b f = \int_a^b f dt$$

It is said that a tagged partition is called γ_ε-fine if satisfies (5).

This definition can be extended on infinite intervals as follows.

Definition 2. *Let* $\gamma : [a, \infty] \rightarrow (0, \infty)$ *be a function, we will say that the tagged partition* $P = \{(\ [t_{i-1}, t_i], s_i)\}_{i=1}^{n+1}$ *is* γ*−fine if:*

(a) $t_0 = a, t_{n+1} = \infty$.

(b) $[t_{i-1}, t_i] \subset [s_i - \gamma_\varepsilon(s_i), s_i + \gamma_\varepsilon(s_i)]$ *for* $i = 1, 2, ..., n$.

(c) $[t_n, \infty] \subset [1/\gamma(\infty), \infty]$.

Put $f(\infty) = 0$ and $f(-\infty) = 0$. This allows us define the integral of f over infinite intervals.

Definition 3. *It is said that the function* $f : [a, \infty] \rightarrow \mathbb{R}$ *is Henstock-Kurzweil integrable if it satisfies the Definition 1, but the partition P must be* γ_ε*−fine according to Definition 2.*

For functions defined on $[-\infty, a]$ or $[-\infty, +\infty]$ the integral is defined analogously. We will denote the vector space of Henstock-Kurzweil integrable functions on I as $HK(I)$

The space of Henstock-Kurzweil integrable functions on the interval $I = [a, b]$, finite or infinite interval, is a semi-normed space with the Alexiewicz semi-norm

$$\|f\|_A = \sup_{a \leq x \leq b} \left| \int_a^x f(t)dt \right|. \tag{6}$$

We denote the space of functions in $HK([c, d])$ for each finite interval $[c, d]$ in I as $HK_{loc}(I)$.

Definition 4. *A function* $f : I \rightarrow \mathbb{R}$ *is a* bounded variation *function over I (finite interval) if exists a* $M > 0$ *such that*

$$\text{Var}(f, I) = \sup \left\{ \sum_{i=1}^{n} |f(t_i) - f(t_{i-1})| : P \text{ is a partition of } I \right\} < M.$$

Its total variation over I is $\text{Var}(f, I)$*. In case I is not finite, for example* $[a, \infty]$*, it is said that* $f : [a, \infty] \rightarrow \mathbb{R}$ *is a bounded variation function over I if there exists* $N > 0$ *such that*

$$\text{Var}(f, [a, t]) \leq N,$$

for all $t \geq a$*. The total variation of f on I is equal to*

$$\text{Var}(f, [a, \infty)) = \sup \left\{ \text{Var}(f, [a, t]) : a \leq t \right\}. \tag{7}$$

For $I = (-\infty, b]$ *the considerations are analogous.*

The set of bounded variation functions over $[a, b]$ is denoted as $BV([a, b])$ and we will denote the space of functions f such that $f \in BV([c, d])$ for each compact interval $[c, d]$ in $\mathbb{R}$ as $BV_{loc}(\mathbb{R})$. We will refer to $BV_0(\mathbb{R})$ as the subspace of functions f belong to $BV(\mathbb{R})$ such that vanishing at $\pm\infty$.

At the Lemma 25 we prove that: $HK(\mathbb{R}) \cap BV(\mathbb{R}) \subset BV_0(\mathbb{R})$. It is not hard prove that $BV_0(\mathbb{R}) \nsubseteq L^1(\mathbb{R})$ and $BV_0(\mathbb{R}) \nsubseteq HK(\mathbb{R})$. Furthermore, there are functions in $HK(\mathbb{R})$ or $L^1(\mathbb{R})$ but they are not in $BV_0(\mathbb{R})$. For example, the function $f(t)$ defined by 0 for $t \in (-\infty, 1)$ and $1/t$ for $t \in [1, \infty)$ belongs to $BV_0(\mathbb{R})$ but does not belong to $L^1(\mathbb{R})$, neither in $HK(\mathbb{R})$. In addition, other examples are the characteristic function of $\mathbb{Q}$ on a compact interval and $g(t) = t^2 \sin(\exp(t^2))$ are in $HK(\mathbb{R}) \setminus BV_0(\mathbb{R})$.

We consider the completion of $HK([a,b])$ as

$$\{[\{f_k\}] : \{f_k\} \text{ is a Cauchy sequence in } HK([a,b])\},$$

where the convergence is respect Alexiewicz norm, and will be denoted by $\widehat{HK([a,b])}$. It is possible to prove that $\widehat{HK([a,b])}$ is isometrically isomorphic to the subspace of distributions each of which is the distributional derivative of a continuous function, see [12]. The indefinite integral of $f = [\{f_k\}] \in \widehat{HK([a,b])}$ is defined as

$$\int_a^x f = \lim_{k \to \infty} \int_a^x f_k.$$

Thus, $\widehat{HK([a,b])}$ is a Banach space with the Alexiewicz norm (6). The completion is also defined in [13]. Besides, basic results of the integral continue being true on the completion. More details see [12].

To facilitate reading, we recall the following results. The first one is a well known result, and it can be found for example in [14] and [15].

Theorem 5. *Let f be a real function defined on $\mathbb{N} \times \mathbb{N}$. If $\lim_{n \to n_0} f(k,n) = \psi(k)$ exists for each k, and $\lim_{k \to k_0} f(k,n) = \varphi(n)$ converges uniformly on n, then*

$$\lim_{k \to k_0} \lim_{n \to n_0} f(k,n) = \lim_{n \to n_0} \lim_{k \to k_0} f(k,n).$$

Theorem 6. *[16, Theorem 33.1] Suppose X is a normed space, Y is a Banach space and that $\{T_n\}$ is a sequence of bounded linear operators from X into Y. Then the conditions: i) $\{\|T_n\|\}$ is bounded and ii) $\{T_n(x)\}$ is convergent for each $x \in Z$, where Z is a dense subset on X implies that for each $x \in X$, the sequence $(T_n(x))$ is convergent in Y and the linear operator $T : X \to Y$ defined by $T(x) = \lim_{n \to \infty} T_n(x)$ is bounded.*

Theorem 7. *[17] If g is a HK integrable function on $[a, b] \subseteq \mathbb{R}$ and f is a bounded variation function on $[a, b]$, then fg is HK integrable on $[a, b]$ and*

$$\left| \int_a^b fg \right| \leq \inf_{t \in [a,b]} |f(t)| \left| \int_a^b g(t)dt \right| + \|g\|_{[a,b]} \text{Var}(f, [a,b]).$$

Theorem 8. *[11, Hake's Theorem]* $\varphi \in HK([a, \infty])$ *if and only if for each* b, ε *such that* $b > a$ $b - a > \varepsilon > 0$, *it follows that* $\varphi \in HK([a + \varepsilon, b])$ *and* $\lim\limits_{\varepsilon \to 0,\, b \to \infty} \int_{a+\varepsilon}^{b} \varphi(t)dt$ *exists. In this case, this limit is* $\int_{a}^{\infty} \varphi(t)dt$.

Theorem 9. *[11, Chartier-Dirichlet's Test] Let* f *and* g *be functions defined on* $[a, \infty)$. *Suppose that*

1. $g \in HK([a, c])$ *for every* $c \geq a$, *and* G *defined by* $G(x) = \int_{a}^{x} g$ *is bounded on* $[a, \infty)$.
2. f *is of bounded variation on* $[a, \infty)$ *and* $\lim\limits_{x \to \infty} f(x) = 0$.

Then $fg \in HK([a, \infty))$.

Moreover, by Multiplier Theorem, Hake's Theorem and Chartier-Dirichlet Test, we have the following lemma.

Lemma 10. *Let* $f, g : [a, \infty] \to \mathbb{R}$. *Suppose that* $f \in BV_0([a, \infty])$, $\varphi \in HK([a, b])$ *for every* $b > a$, *and* $\Phi(t) = \int_{a}^{t} \varphi du$ *is bounded on* $[a, \infty)$. *Then* $\varphi f \in HK([a, b])$,

$$\int_{a}^{\infty} \varphi f \, dt = -\int_{a}^{\infty} \Phi(t) df(t)$$

and

$$\left| \int_{a}^{\infty} \varphi f \, dt \right| \leq \sup_{a < t} |\Phi(t)| \operatorname{Var}(f, [a, \infty]).$$

Similar results are valid for the cases $[-\infty, \infty]$ *and* $[-\infty, a]$.

Let $I = [a, b]$ and $E \subset I$. We say that the function $F : I \to \mathbb{R}$ is in $AC_{\delta}(E)$ if for each $\epsilon > 0$ there exist $\eta_{\varepsilon} > 0$ and a gauge δ_{ε} on E such that if $\{(x_i, y_i)\}_i^{N}$ is a $(\delta_{\varepsilon}, E)$−fine subpartition of E such that $\sum_i^{N} (y_i - x_i) < \eta_{\varepsilon}$, then $\sum_1^{N} |F(x_i) - F(y_i)| < \epsilon$. On the other hand, F belongs to the class $ACG_{\delta}(I)$ if there exists a sequence $\{E_n\}_1^{\infty}$ of sets in I such that $I = \cup_{n=1}^{\infty} E_n$ and $F \in AC_{\delta}(E_n)$ for each $n \in \mathbb{N}$. A characterization of this type of functions is the following.

Theorem 11. *A function* $f \in HK(I)$ *if and only if there exists a function* $F \in ACG_{\delta}$ *such that* $F' = f$ *a.e.*

Theorem 12. *[18, Theorem 4] Let* $a, b \in \mathbb{R}$. *If* $h : \mathbb{R} \times [a, b] \to \mathbb{C}$ *is such that*

1. $h(t, \cdot)$ *belongs to* ACG_{δ} *on* $[a, b]$ *for almost all* $t \in \mathbb{R}$;
2. *and* $h(\cdot, s)$ *is a HK integrable function on* $\mathbb{R}$ *for all* $s \in [a, b]$.

Then $H := \int_{-\infty}^{\infty} h(t, \cdot)dt$ *belongs to* ACG_{δ} *on* $[a, b]$ *and* $H'(s) = \int_{-\infty}^{\infty} D_2 h(t, s)dt$ *for almost all* $s \in (a, b)$, *iff,*

$$\int_{s}^{t} \int_{-\infty}^{\infty} D_2 h(t, s)dt ds = \int_{-\infty}^{\infty} \int_{s}^{t} D_2 h(t, s)ds dt$$

for all $[s, t] \subseteq [a, b]$. *In particular,*

$$H'(s_0) = \int_{-\infty}^{\infty} D_2 h(t, s_0) dt$$

when $H_2 := \int_{-\infty}^{\infty} D_2 h(t, \cdot) dt$ *is continuous at* s_0.

2. Fourier coefficients for functions in the Henstock-Kurzweil completion.

For finite intervals, the Theorem 12.11 of [19] tells us that: In order that $\int_a^b f g_n \to \int_a^b f g$, $n \to \infty$, whenever $f \in HK([a,b])$, it is necessary and sufficient that: i) g_n is almost everywhere of bounded variation on $[a, b]$ for each n; ii) $\sup\{||g_n||_\infty + ||g_n||_{BV}\} < \infty$; iii) $\int_c^d g_n \to \int_c^d g$, $n \to \infty$, for each interval $(c, d) \subset (a, b)$. The Theorem 3 of [8] proves that above theorem is valid for infinite intervals. In this section we show that [19, Theorem 12.11] and [8, Theorem 3] are true for functions belonging to the completion of the Henstock-Kurzweil space. First, we need to prove the next lemma. The class of step functions on $[a, b]$ will be denoted as $K([a, b])$.

Lemma 13. *Let $[a, b]$ be an infinite interval. The set $K([a, b])$ is dense in $HK([a, b])$.*

Proof. Let $f \in HK([a, \infty])$ and $\epsilon > 0$ be given. By Hake's Theorem, exists $N \in \mathbb{N}$ such that for each $x \geq N$,

$$\left| \int_x^\infty f \right| < \frac{\epsilon}{2}. \tag{8}$$

Since $K([a, N])$ is dense in $HK([a, N])$, by Theorem 7 of [20], there exists a function $h \in K([a, N])$ such that

$$||f - h||_{A,[a,N]} = \sup_{x \in [a,N]} \left| \int_a^x (f - h) \right| < \frac{\epsilon}{2}. \tag{9}$$

Defining $h_0 \in K([a, \infty])$ as

$$h_0(x) = \begin{cases} h(x) \text{ if } x \in [a, N] \\ 0 \text{ if } x \in (N, \infty]. \end{cases}$$

It follows, by (8) and (9), that

$$||f - h_0||_A \leq \epsilon.$$

Similar arguments apply for intervals as $[-\infty, a]$ or $[-\infty, \infty]$. $\square$

2.1. The convergence of integrals of products in the completion

The following result appears in [3]. Here, we present a detailed proof.

Theorem 14. *Let* $[a,b] \subset \overline{\mathbb{R}}$. *In order that*

$$\int_a^b fg_n \to \int_a^b fg, \quad n \to \infty, \tag{10}$$

whenever $f \in \widehat{HK([a,b])}$, *it is necessary and sufficient that: i)* g_n *is almost everywhere of bounded variation on* $[a,b]$ *for each* n; *ii)* $\sup\{\|g_n\|_\infty + \|g_n\|_{BV}\} < \infty$; *iii)* $\int_c^d g_n \to \int_c^d g$, $n \to \infty$, *for each interval* $(c,d) \subset (a,b)$.

Proof. The necessity follows from [19, Theorem 12.11]. Now we will prove the sufficiency condition. Define the linear functionals $T, T_n : \widehat{HK([a,b])} \to \mathbb{R}$ by

$$T_n(f) = \int_a^b fg_n \quad \text{and} \quad T(f) = \int_a^b fg. \tag{11}$$

Supposing *i)* and *ii)*, we have, by Multiplier Theorem, that the sequence $\{T_n\}$ is bounded by $\sup\{\|g_n\|_\infty + \|g_n\|_{BV}\}$. Owing to Lemma 13, the space of step functions is dense in $\widehat{HK([a,b])}$, then considering the Theorem 6 it is sufficient to prove that $\{T_n(f)\}$ converge to $T(f)$, for each step function f. First, let $f(x) = \chi_{(c,d)}(x)$ be the characteristic function of $(c,d) \subset [a,b]$. Thus,

$$T_n(f) = \int_a^b \chi_{(c,d)} g_n = \int_c^d g_n,$$

by the hypothesis *iii)*, we have that $\{T_n(\chi_{(c,d)})\}$ converges to $T(\chi_{(c,d)})$, as $n \to \infty$. Now, let f be a step function. Being that each T_n is a linear functional, then $\{T_n(f)\}$ converges to $T(f)$, as $n \to \infty$. Thus, the result holds.   □

Remark 15. *On* $HK([a,b])$. *The hypothesis iii) can be replaced by:* g_n *converges pointwise to* g, *then the result follows from Corollary 3.2 of [21]. The result on the completion holds by Theorem 6. Note that the conditions i), ii) and iii) do not imply converges pointwise from* $\{g_n\}$ *to* g, *see example 2 of [8].*

For the case of functions defined on a finite interval we get Theorem 16, and a lemma of Riemann-Lebesgue type for functions in the Henstock-Kurzweil space completion.

Theorem 16. *Let* $[a,b]$ *be a finite interval. If i)* g_n *converges to* g *in measure on* $[a,b]$, *ii) each* g_n *is equal to* h_n *almost everywhere, a normalized bounded variation function and iii) there is* $M > 0$ *such that* $Var(h_n, [a,b]) \leq M$, $n \geq 1$, *then for all* $f \in \widehat{HK([a,b])}$,

$$\int_a^b fg_n \to \int_a^b fg, \quad n \to \infty.$$

Proof. Let $f \in \widehat{HK([a,b])}$ be given, where $f = [\{f_k\}]$, we want to prove that

$$\lim_{n \to \infty} \lim_{k \to \infty} \int_a^b f_k g_n = \int_a^b fg.$$

Define $f(k,n) = \int_a^b f_k g_n$. By the hypothesis *i)* about (g_n) we have

$$\lim_{n \to \infty} \int_a^b f_k g_n = \int_a^b f_k g.$$

Moreover

$$\lim_{k \to \infty} \lim_{n \to \infty} \int_a^b f_k g_n = \int_a^b fg,$$

by the integral definition on the completion. We will prove that $\lim_{k \to \infty} f(k,n) = \int_a^b fg_n$ converges uniformly on n. Let $\epsilon > 0$ be given, there exists k_0 such that $||f_k - f||_A \leq \epsilon$, if $k \geq k_0$. Besides, if $k \geq k_0$,

$$\left| \int_a^b f_k g_n - \int_a^b fg_n \right| \leq ||f_k - f||_A Var(g_n, [a,b])$$

$$\leq M\epsilon.$$

Therefore, by Theorem 5,

$$\lim_{n \to \infty} \lim_{k \to \infty} \int_a^b f_k g_n = \int_a^b fg.$$

$\square$

The following result is a "generalization" of Riemann-Lebesgue Lemma on the completion of the space $HK([a,b])$, over finite intervals, it also appears in [3].

Corollary 17. *If $\varphi : \mathbb{R} \to \mathbb{R}$ such that φ' exists, is bounded and $\varphi(s) = o(s)$, as $|s| \to \infty$, then for each $f \in \widehat{HK([a,b])}$ we have the next asymptotic behavior*

$$\int_a^b \varphi(st) f(t) dt = o(s), \quad as \ |s| \to \infty.$$

Proof. For each $s \neq 0$ define $\varphi_s : \mathbb{R} \to \mathbb{R}$ as $\varphi_s(t) = \varphi(st)/s$. In order to prove

$$\lim_{s \to \infty} \int_a^b \frac{\varphi(st)}{s} f(t) dt = 0,$$

it is sufficient to show that φ_s fulfills the hypothesis of Theorem 16. Now, we will check item by item. *i)* Because of $\varphi(s) = o(s)$, as $|s| \to \infty$ and the interval $[a,b]$ is finite, it follows that φ_s converges in measure to 0. *ii)* Owing to φ' is bounded then, by the Mean Value Theorem, we have that $\varphi_s \in BV([a,b])$. *iii)* $Var(\varphi_s, [a,b])$ is bounded uniformly by upper bound of φ' and $a - b$.

$\square$

2.2. Riemann-Lebesgue Property

This property establishes that $\int_r^\pi f(t)D_n(t)dt \to 0$, for each $f \in L^1[-\pi, \pi]$ and $r \in (0, \pi]$, where $D_n(t) = \frac{\sin(n+1/2)t}{\sin(t/2)}$ denotes the n-th Dirichlet Kernel of order n. Now, we provide an analogous result concerning the Henstock-Kurzweil completion.

Theorem 18. *For any $f \in \widehat{HK([-\pi, \pi])}$, and $r \in (0, \pi]$,*

$$\lim_{n \to \infty} \frac{1}{n} \int_r^\pi f(t)D_n(t)dt = 0. \tag{12}$$

Proof. Note that the function $g(t) = 1/\sin(t/2)$ is in $BV([r, \pi])$. Moreover, by Multiplier Theorem we have $fg \in \widehat{HK([r, \pi])}$. Hence, by Corollary 17, we get

$$\int_r^\pi f(t)g(t)\sin(n + 1/2)tdt = o(n), \quad |n| \to \infty.$$

$\square$

Considering an similar argument from above proof, it follows that

$$\int_r^\pi f(t)\frac{\sin(n + 1/2)t}{t/2}dt = o(n), \quad |n| \to \infty. \tag{13}$$

For $n \in \mathbb{N} \cup \{0\}$, we define the function $\Phi_n(t) = \frac{\sin(n+1/2)t}{t/2}$ for $t \neq 0$ and $\Phi_n(0) = 2n + 1$, it is called the discrete Fourier Kernel of order n. This kernel provides a very good approximation to the Dirichlet Kernel D_n for $|t| < 2$, but Φ_n decreases more rapidly than D_n, see [1].

Theorem 19. *Let $f \in \widehat{HK([0, \pi])}$ and $r \in (0, \pi]$. Then, assuming that any of next limits exist,*

$$\lim_{n \to \infty} \frac{1}{n} \int_0^r f(t)D_n(t)dt = \lim_{n \to \infty} \frac{1}{n} \int_0^r f(t)\frac{\sin(n + 1/2)t}{t/2}dt.$$

Proof. Define $g : [0, \pi] \to \mathbb{R}$ by

$$g(t) = \begin{cases} \frac{1}{\sin(t/2)} - \frac{1}{t/2} & \text{for } t \in (0, \pi] \\ \\ 0 & \text{for } t = 0. \end{cases}$$

Since $g \in BV([0, \pi])$, $fg \in \widehat{HK([0, \pi])}$. By Corollary 17, we have

$$\lim_{n\to\infty} \frac{1}{n} \int_0^\pi f(t) \left(\frac{1}{\sin(t/2)} - \frac{1}{t/2} \right) \sin(n+1/2)\, t\, dt = 0. \tag{14}$$

Now, by (14), we have

$$\lim_{n\to\infty} \frac{1}{n} \int_0^\pi \left(f(t)D_n(t) - f(t)\frac{\sin(n+1/2)t}{t/2} \right) dt = 0.$$

Then

$$\lim_{n\to\infty} \frac{1}{n} \left[\int_0^r \left(f(t)D_n(t) - f(t)\frac{\sin(n+1/2)t}{t/2} \right) dt \right.$$
$$\left. + \int_r^\pi \left(f(t)D_n(t) - f(t)\frac{\sin(n+1/2)t}{t/2} \right) dt \right] = 0.$$

By Theorem 18 and (13),

$$\lim_{n\to\infty} \frac{1}{n} \int_r^\pi \left(f(t)D_n(t) - f(t)\frac{\sin(n+1/2)t}{t/2} \right) dt = 0.$$

Therefore, assuming that any of the limits exist, we have

$$\lim_{n\to\infty} \frac{1}{n} \int_0^r f(t)D_n(t)\, dt = \lim_{n\to\infty} \frac{1}{n} \int_0^r f(t)\frac{\sin(n+1/2)t}{t/2}\, dt.$$

$\square$

The following result is a characterization of the asymptotic behavior of $n-th$ partial sum of the Fourier series, it can be found in [3].

Corollary 20. *Let $f \in \widehat{HK([-\pi,\pi])}$ be $2\pi-$ periodic. The $n-th$ partial sum of the Fourier series at t has the following asymptotic behavior $S_n(f,t) = o(n)$, when $|n| \to \infty$ iff*

$$\int_0^\pi [f(t+u) + f(t-u)]\frac{\sin(n+1/2)u}{u}\, du = o(n),$$

if $|n| \to \infty$.

Proof. Since $S_n(f,t) = \int_{-\pi}^\pi f(t+u)D_n(u)\, du$, realizing a change of variable (see section 6 of [13]), then by Theorem 19 we get the result. $\square$

3. Henstock-Kurzweil integral transform

The results in this section are based for functions in the vector space $BV_0(\mathbb{R})$, and they have to [10] as principal reference.

We will introduce some additional terminology in order to facilitate the following results.

If $g : \mathbb{R} \times \mathbb{R} \to \mathbb{C}$ is a function and $s_0 \in \mathbb{R}$, we say that s_0 fulfills hypothesis **(H)** relative to g if:

(H) there exist $\delta = \delta(s_0) > 0$ and $M = M(s_0) > 0$, such that, if $|s - s_0| < \delta$ then

$$\left| \int_u^v g(t,s)dt \right| \leq M,$$

for all $[u, v] \subseteq \mathbb{R}$.

This condition plays a significant role in the following results. Also, the next theorems can be found in [10].

Theorem 21. *Let $f : \mathbb{R} \to \mathbb{R}$ and $g : \mathbb{R} \times \mathbb{R} \to \mathbb{C}$ be functions. Assume that $f \in BV_0(\mathbb{R})$, and $s_0 \in \mathbb{R}$ fulfills Hypothesis **(H)** relative to g, then*

$$\Gamma(s) = \int_{-\infty}^{\infty} f(t)g(t,s)dt$$

is well defined for all s in a neighborhood of s_0.

Proof. Applying Theorem 9 the result holds.   □

Theorem 22. *Let $f : \mathbb{R} \to \mathbb{R}$ and $g : \mathbb{R} \times \mathbb{R} \to \mathbb{C}$ be functions assume that*

1. *f belongs to $BV_0(\mathbb{R})$, g is bounded, and*
2. *$g(t, \cdot)$ is continuous for all $t \in \mathbb{R}$.*

*If $s_0 \in \mathbb{R}$ fulfills Hypothesis **(H)** relative to g, then the function Γ is continuous at s_0.*

Proof. By Hypothesis **(H)**, there exist $\delta_1 > 0$ and $M > 0$, such that, if $|s - s_0| < \delta_1$ then

$$\left| \int_u^v g(t,s)dt \right| \leq M \qquad (15)$$

for all $[u, v] \subseteq \mathbb{R}$. From Theorem 21, $\Gamma(s)$ exists for all $s \in B_{\delta_1}(s_0)$.

Let an arbitrary $\epsilon > 0$, by Hake's theorem, there exists $K_1 > 0$ such that

$$\left| \int_{|t| \geq u} f(t)g(t, s_0)dt \right| < \frac{\epsilon}{3} \tag{16}$$

for all $u \geq K_1$. On the other hand, as

$$\lim_{t \to -\infty} \text{Var}(f, (-\infty, t]) = 0 \quad \text{and} \quad \lim_{t \to \infty} \text{Var}(f, [t, \infty)) = 0,$$

there is $K_2 > 0$ such that for each $t > K_2$,

$$\text{Var}(f, (-\infty, -t]) + \text{Var}(f, [t, \infty)) < \frac{\epsilon}{3M}.$$

Let $K = \max\{K_1, K_2\}$. From Theorem 7, it follows that for every $v \geq K$ and every $s \in B_{\delta_1}(s_0)$,

$$\left| \int_K^v f(t)g(t, s)dt \right| \leq \|g(\cdot, s)\|_{[K, v]} \left[\inf_{t \in [K, v]} |f(t)| + \text{Var}(f, [K, v]) \right]$$
$$\leq M \left[|f(v)| + \text{Var}(f, [K, \infty)) \right],$$

where the second inequality is true due to (15). This implies, since $\lim_{t \to \infty} |f(t)| = 0$, that

$$\left| \int_K^\infty f(t)g(t, s)dt \right| \leq M \cdot \text{Var}(f, [K, \infty)).$$

Analogously we have that

$$\left| \int_{-\infty}^{-K} f(t)g(t, s)dt \right| \leq M \cdot \text{Var}(f, (-\infty, -K]).$$

Therefore, for each $s \in B_{\delta_1}(s_0)$,

$$\left| \int_{|t| \geq K} f(t)g(t, s)dt \right| \leq M \left[\text{Var}(f, (-\infty, -K])f + \text{Var}(f, [K, \infty)) \right]$$
$$< M \frac{\epsilon}{3M} = \frac{\epsilon}{3}. \tag{17}$$

Since f is $L^1[-K, K]$, g is bounded and $g(t, \cdot)$ is continuous for all $t \in \mathbb{R}$. For example, using Theorem 12.12 of [11], it is easy to show that the function

$$\Gamma_K(s) = \int_{-K}^{K} f(t)g(t, s)dt, \quad s \in \mathbb{R},$$

is continuous at s_0. This implies that there is $\delta_2 > 0$ such that for every $s \in B_{\delta_2}(s_0)$,

$$\left| \int_{-K}^{K} f(t)[g(t, s) - g(t, s_0)]dt \right| < \frac{\epsilon}{3}. \tag{18}$$

Let $\delta = \min\{\delta_1, \delta_2\}$. Then for all $s \in B_\delta(s_0)$,

$$|\Gamma(s) - \Gamma(s_0)| \leq \left| \int_{-K}^{K} f(t)[g(t, s) - g(t, s_0)]dt \right|$$
$$+ \left| \int_{|t| \geq K} f(t)g(t, s)dt \right| + \left| \int_{|t| \geq K} f(t)g(t, s_0)dt \right|.$$

Thus, from (16), (17) and (18), $|\Gamma(s) - \Gamma(s_0)| < \frac{\epsilon}{3} + \frac{\epsilon}{3} + \frac{\epsilon}{3} = \epsilon$, for all $s \in B_\delta(s_0)$. $\square$

Theorem 23. *Let $a, b \in \mathbb{R}$. If $f : \mathbb{R} \to \mathbb{R}$ and $g : \mathbb{R} \times [a, b] \to \mathbb{C}$ are functions such that*

1. *$f \in BV_0(\mathbb{R})$, g is measurable, bounded and*
2. *for all $s \in [a, b]$, s satisfies Hypothesis **(H)** relative to g.*

Then

$$\int_{a}^{b} \int_{-\infty}^{\infty} f(t)g(t, s)dt\,ds = \int_{-\infty}^{\infty} \int_{a}^{b} f(t)g(t, s)ds\,dt$$

Proof. From (2) and since $[a, b]$ is compact, there exists $M > 0$ such that, for every $s \in [a, b]$ and for all $[u, v] \subseteq \mathbb{R} : \left| \int_u^v g(t, s)dt \right| \leq M$.

For $r > 0$ and $s \in [a, b]$, let $\Gamma_r(s) = \int_{-r}^{r} f(t)g(t, s)dt$. By Theorem 7, we notice that

$$|\Gamma_r(s)| = \left| \int_{-r}^{r} f(t)g(t, s)dt \right|$$
$$\leq \|g(\cdot, s)\|_{[-r, r]} \left[\inf_{t \in [-r, r]} |f(t)| + V_{[-r, r]}f \right]$$
$$\leq M[|f(0)| + Vf]$$

for all $s \in [a, b]$.

So, for each $r > 0$, Γ_r is HK integrable on $[a, b]$ and is bounded for a fixed constant. Moreover, by Theorem 21 and Hake's theorem

$$\lim_{r \to \infty} \Gamma_r(s) = \Gamma(s)$$

for all $s \in [a, b]$.

Using the Lebesgue Dominated Convergence Theorem, we have that Γ is HK integrable on $[a, b]$ and

$$\int_a^b \Gamma(s)ds = \lim_{r \to \infty} \int_a^b \Gamma_r(s)ds.$$

Now, because of f is Lebesgue integrable on $[-r, r]$; g is measurable and bounded; and by Fubini's theorem, it follows that

$$\int_a^b \int_{-r}^r f(t)g(t,s)dtds = \int_{-r}^r \int_a^b f(t)g(t,s)\,dsdt.$$

Consequently

$$\lim_{r \to \infty} \int_{-r}^r \int_a^b f(t)g(t,s)dsdt = \lim_{r \to \infty} \int_a^b \Gamma_r(s)ds = \int_a^b \Gamma(s)ds.$$

So by Hake's theorem,

$$\int_{-\infty}^\infty \int_a^b f(t)g(t,s)\,dsdt = \int_a^b \Gamma(s)ds = \int_a^b \int_{-\infty}^\infty f(t)g(t,s)dtds.$$

$\square$

Theorem 24. *Let $f \in BV_0(\mathbb{R})$ and $g : \mathbb{R} \times \mathbb{R} \to \mathbb{C}$ be a function such that its partial derivative D_2g is bounded and continuous on $\mathbb{R} \times \mathbb{R}$. If $s_0 \in \mathbb{R}$ is such that*

1. *there is $K > 0$ for which $\|g(\cdot, s_0)\|_{[u,v]} \le K$ for all $[u, v] \subseteq \mathbb{R}$, and*
2. *s_0 satisfies Hypothesis (**H**) relative to D_2g.*

Then Γ is derivable at s_0, and

$$\Gamma'(s_0) = \int_{-\infty}^\infty f(t)D_2g(t, s_0)dt. \tag{19}$$

Proof. Using conditions (1) and (2) and the Mean Value theorem, there exist $\delta > 0$ and $M > 0$ such that, for each $s \in (s_0 - \delta, s_0 + \delta)$,

$$\left| \int_u^v D_2 g(t,s)dt \right| < M \quad \text{and} \quad \left| \int_u^v g(t,s)dt \right| < M, \tag{20}$$

for all $[u,v] \subseteq \mathbb{R}$.

Let a, b be real numbers with $s_0 - \delta < a < s_0 < b < s_0 + \delta$. We use Theorem 12 to prove (19). The function $f(t)g(t,\cdot)$ is differentiable on $[a,b]$ for each $t \in \mathbb{R}$, therefore $f(t)g(t,\cdot)$ is ACG_δ on $[a,b]$ for all $t \in \mathbb{R}$. By (20) and Theorem 9, $f(\cdot)g(\cdot,s)$ is HK-integrable on $\mathbb{R}$ for all $s \in [a,b]$. Then

$$\Gamma'(s_0) = \int_{-\infty}^{\infty} f(t)D_2 g(t,s_0)dt$$

when, if

$$\Gamma_2 := \int_{-\infty}^{\infty} f(t)D_2 g(t,\cdot)dt$$

is continuous at s_0, and

$$\int_s^t \int_{-\infty}^{\infty} f(t)D_2 g(t,s)dtds = \int_{-\infty}^{\infty} \int_s^t f(t)D_2 g(t,s)dsdt$$

for all $[s,t] \subseteq [a,b]$. The first affirmation is true by (20) and Theorem 22, and the second affirmation is true due to (20) and Theorem 23   □

3.1. Some applications

An important work about the Fourier transform using the Henstock-Kurzweil integral: existence, continuity, inversion theorems etc. was published in [5]. Nevertheless, there are some omissions in that results that use the Lemma 25 (a) of [5]. Also the authors of this book chapter in [6], [3] and [4] have studied existence, continuity and Riemann-Lebesgue lemma about the Fourier transform of functions belong to $HK(\mathbb{R}) \cap BV(\mathbb{R})$ and $BV_0(\mathbb{R})$. Following the line of [6], in Theorem 26 we include some results from them as consequences of theorems above section.

Let f and g be real-valued functions on $\mathbb{R}$. The convolution of f and g is the function $f * g$ defined by

$$f * g(x) = \int_{-\infty}^{\infty} f(x - y)g(y)dy$$

for all x such that the integral exists. Several conditions can be imposed on f and g to guarantee that $f * g$ is defined on $\mathbb{R}$. For example, if f is HK- integrable and g is of bounded variation.

Lemma 25. *For $f \in HK(\mathbb{R}) \cap BV(\mathbb{R})$, $\lim\limits_{|x| \to \infty} f(x) = 0$.*

Proof. Since f is a bounded variation function on $\mathbb{R}$ then the limit of $f(x)$, as $|x| \to \infty$, exists. Suppose that $\lim\limits_{|x| \to \infty} f(x) = \alpha \neq 0$. Take $0 < \epsilon < |\alpha|$. There exists $A > 0$ such that $\alpha - \epsilon < f(x)$, for all $|x| > A$. Observe that $f(x) > 0$ on $[A, \infty)$, so $f \in L([A, \infty))$. Therefore the constant function $\alpha - \epsilon$ is Lebesgue integrable on $[A, \infty)$, which is a contradiction. $\square$

Observe, as consequence of above Lemma, we have that the vector space $HK(\mathbb{R}) \cap BV(\mathbb{R})$ is contained in $BV_0(\mathbb{R})$. So the next theorem is an immediately consequence of above section.

Theorem 26. *If $f \in HK(\mathbb{R}) \cap BV(\mathbb{R})$, then*

1. *$\widehat{f}$ exists on $\mathbb{R}$.*

2. *$\widehat{f}$ is continuous on $\mathbb{R} \setminus \{0\}$.*

3. *If $g(t) = tf(t)$ and $g \in HK(\mathbb{R}) \cap BV(\mathbb{R})$, then $\widehat{f}$ is differentiable on $\mathbb{R} \setminus \{0\}$, and*

$$\widehat{f}'(s) = -i\widehat{g}(s), \quad \text{for each } s \in \mathbb{R} \setminus \{0\}.$$

4. *For $h \in L^1(\mathbb{R}) \cap BV(\mathbb{R})$, $\widehat{f * h}(s) = \widehat{f}(s)\widehat{h}(s)$ for all $s \in \mathbb{R}$.*

Proof. We observe that

$$\left| \int_u^v e^{-its} dt \right| \leq \frac{2}{|s|}, \tag{21}$$

for all $[u, v] \subseteq \mathbb{R}$. Thus, each $s_0 \neq 0$ satisfies Hypothesis (**H**) relative to e^{-its}.

(a) Theorem 21 implies that $\widehat{f}(s_0)$ exists for all $s_0 \neq 0$ and, since $f \in HK(\mathbb{R})$, $\widehat{f}(0)$ exists. Therefore $\widehat{f}$ exists on $\mathbb{R}$.

(b) By Theorem 22, $\widehat{f}$ is continuous at s_0, for all $s_0 \neq 0$.

(c) It follows by Theorem 12 in similar way to the proof of Theorem 24.

(d) Let $k(x, y) = f(y - x)e^{-iys}$, where s is a fixed real number. We get, for each $y \in \mathbb{R}$ and all $[u, v] \subseteq \mathbb{R}$,

$$\left| \int_u^v k(x, y) dx \right| = \left| \int_u^v f(y - x) dx \right|$$

$$= \left| \int_{y-u}^{y-v} f(z) dz \right| \leq \|f\|_A.$$

So, every real number y satisfies Hypothesis (**H**) relative to k. Now, observe that $h \in BV_0(\mathbb{R})$ and k is measurable and bounded. Thus, by Theorem 23,

$$\int_{-a}^{a} \int_{-\infty}^{\infty} h(x)k(x,y)dxdy = \int_{-\infty}^{\infty} \int_{-a}^{a} h(x)k(x,y)dydx, \tag{22}$$

for all $a > 0$.

On the other hand,

$$\left| h(x) \int_{-a}^{a} f(y-x)e^{-iys}dy \right| \le |h(x)| \left| \int_{-a-x}^{a-x} f(z)e^{-izs}dz \right|$$

$$\le |h(x)| \|f(\cdot)e^{-i(\cdot)s}\|_A.$$

Since $h \in L(\mathbb{R})$, using Dominated Convergence theorem, it follows that

$$\widehat{f}(s)\widehat{h}(s) = \int_{-\infty}^{\infty} h(x) \int_{-\infty}^{\infty} f(y-x)e^{-iys}dydx$$

$$= \lim_{a \to \infty} \int_{-\infty}^{\infty} h(x) \int_{-a}^{a} f(y-x)e^{-iys}dydx.$$

Moreover, from (22), we have

$$\widehat{f}(s)\widehat{h}(s) = \lim_{a \to \infty} \int_{-a}^{a} \int_{-\infty}^{\infty} h(x)f(y-x)e^{-iys}dxdy$$

$$= \lim_{a \to \infty} \int_{-a}^{a} (f*h)(y)e^{-iys}dy.$$

We conclude, by Hake's theorem, that

$$\widehat{f*h}(s) = \widehat{f}(s)\widehat{h}(s).$$

$\square$

Recall that the Laplace transform, at $z \in \mathbb{C}$, of a function $f : [0, \infty) \to \mathbb{R}$ is defined as

$$L(f)(z) = \int_{0}^{\infty} f(t)e^{-zt}dt.$$

Theorem 27. *If $f \in HK([0,\infty)) \cap BV([0,\infty))$, then*

1. $L(f)(z)$ *exists for all $z \in \mathbb{C}$.*
2. *If $F(x,y) = L(f)(x+iy)$, then $F(\cdot, y)$ is continuous on $\mathbb{R}$ for all $y \ne 0$, and $F(x, \cdot)$ is continuous on $\mathbb{R}$ for all $x \ne 0$.*

4. A set of functions in $HK(\mathbb{R}) \cap BV_0(\mathbb{R}) \setminus L^1(\mathbb{R})$

Taking into account Lemma 25, the set $HK(\mathbb{R}) \cap BV(\mathbb{R})$ is included in $BV_0(\mathbb{R})$ and does not have inclusion relations with $L^1(\mathbb{R})$. Since the step functions belong to $HK(\mathbb{R}) \cap BV(\mathbb{R})$, then by Lemma 13, we have that $HK(\mathbb{R}) \cap BV(\mathbb{R})$ is dense in $HK(\mathbb{R})$. In this section we exhibit a set of functions in $HK(\mathbb{R}) \cap BV_0(\mathbb{R}) \setminus L^1(\mathbb{R})$.

Proposition 28. *Let $b > a > 0$. Suppose that $f : [a, \infty) \to \mathbb{R}$ is not identically zero, is continuous and periodic with period $b - a$. Let $F(x) = \int_a^x f(t)dt$ be bounded on $[a, \infty)$. Moreover, assume that $\varphi : [a, \infty) \to \mathbb{R}$ is a nonnegative and monotone decreasing function which satisfies the next conditions:*

(i) $\lim_{t \to \infty} \varphi(t) = 0$,

(ii) $\varphi \notin HK([a, \infty))$.

Then the product $\varphi f \in HK([a, \infty)) \setminus L^1([a, \infty))$.

Proof. We take $t_o \in (a, b)$, $\delta_o > 0$ and $\gamma > 0$ such that

$$\gamma \leq |f(t)| \quad \text{for each} \quad t \in [t_o - \delta_o, t_o + \delta_o] \subset (a, b).$$

Periodicity of f gives

$$\gamma \leq |f(t)|$$

for each $t \in \bigcup_{k=0}^{\infty}[t_o - \delta_o + k(b - a), t_o + \delta_o + k(b - a)]$. Therefore,

$$\int_a^{b+n(b-a)} \varphi(t)|f(t)|dt \geq \gamma \sum_{k=0}^{n} \int_{t_o-\delta_o+k(b-a)}^{t_o+\delta_o+k(b-a)} \varphi(t)dt$$

$$\geq \gamma \sum_{k=0}^{n} \int_{t_o-\delta_o+k(b-a)}^{t_o+\delta_o+k(b-a)} \varphi(t_o + \delta_o + k(b - a))dt$$

$$= \gamma(2\delta_o) \sum_{k=0}^{n} \varphi(t_o + \delta_o + k(b - a)). \tag{23}$$

Also,

$$\int_a^{b+n(b-a)} \varphi(t)dt \leq \sum_{k=0}^{n} \int_{a+k(b-a)}^{b+k(b-a)} \varphi(t)dt$$

$$\leq \sum_{k=0}^{n} \varphi(a + k(b - a)) \int_{a+k(b-a)}^{b+k(b-a)} dt$$

$$\leq (b - a)\varphi(a) \tag{24}$$

$$+ (b - a) \sum_{k=1}^{n} \varphi(t_o + \delta_o + (k - 1)(b - a)).$$

Because of $\varphi \notin HK([a,\infty])$, we get $\lim_{n\to\infty} \int_a^{b+n(b-a)} \varphi(t)dt = \infty$. Thus, equations (23) and (24) imply $\varphi f \notin L^1([a,\infty)]$. On the other hand, by Chartier-Dirichlet's Test of [11], the function φf belongs to $HK([a,\infty))$. □

Corollary 29. *Let α, β be positive numbers such that $\alpha + \beta > 1$ with $\beta \leq 1$. Suppose $a > 0$ and $f : [a,\infty) \to \mathbb{R}$ obeys the hypotheses of Proposition 28. Then, the function $f_{\alpha,\beta} : [a^{1/\alpha},\infty) \to \mathbb{R}$ defined by*

$$f_{\alpha,\beta}(t) = \frac{f(t^\alpha)}{t^\beta} \tag{25}$$

is in $HK([a^{1/\alpha},\infty)) \setminus L^1([a^{1/\alpha},\infty))$.

Proof. The change of variable $u = t^\alpha$ gives,

$$\int_{a^{\frac{1}{\alpha}}}^\infty \frac{f(t^\alpha)}{t^\beta}dt = \int_a^\infty \frac{f(u)}{u^{\frac{\beta-1}{\alpha}+1}}du. \tag{26}$$

The hypotheses for α, β imply that the function $\varphi(u) = u^{-[\frac{(\beta-1)}{\alpha}+1]}$ satisfies the conditions of Proposition 28. Then, $\varphi f \in HK([a,\infty) \setminus L^1([a,\infty))$, satisfying the statement of the corollary. □

Proposition 30. *Let $\beta > \alpha > 0$ be fixed with $\beta + \alpha > 1$. Suppose $f : [a,\infty) \to \mathbb{R}$ is a bounded and continuous function, with bounded derivative. Then the function $f_{\alpha,\beta} : [a^{1/\alpha},\infty) \to \mathbb{R}$, defined by $f_{\alpha,\beta}(t) = f(t^\alpha)/t^\beta$, belongs to the space $BV([a^{1/\alpha},\infty))$.*

Proof. Let M_1 and M_2 be bounds for f and f', respectively. We have,

$$f'_{\alpha,\beta}(t) = \frac{\alpha f'(t^\alpha)}{t^{\beta-\alpha+1}} - \frac{\beta f(t^\alpha)}{t^{\beta+1}},$$

which gives

$$\left| f'_{\alpha,\beta}(t) \right| \leq \frac{\alpha M_2}{t^{\beta-\alpha+1}} + \frac{\beta M_1}{t^{\beta+1}}.$$

Now, take $x > a^{\frac{1}{\alpha}}$. Since $\beta - \alpha > 0$, then

$$\frac{\alpha M_2}{t^{\beta-\alpha+1}} + \frac{\beta M_1}{t^{\beta+1}} \in L^1([a^{\frac{1}{\alpha}}, x)).$$

A straightforward application of the Theorem 7.7 of [11] implies $f'_{\alpha,\beta} \in L^1([a^{1/\alpha}, x))$. Moreover

$$\int_{a^{\frac{1}{\alpha}}}^{x} |f'_{\alpha,\beta}(t)| dt \leq \alpha M_2 \int_{a^{\frac{1}{\alpha}}}^{x} t^{-\beta+\alpha-1} dt$$

$$+ \beta M_1 \int_{a^{\frac{1}{\alpha}}}^{x} t^{-\beta-1} dt$$

$$= -\frac{\alpha M_2}{\beta - \alpha} \left(\frac{1}{x^{\beta-\alpha}} - \frac{1}{a^{\frac{\beta-\alpha}{\alpha}}} \right)$$

$$- M_1 \left(\frac{1}{x^\beta} - \frac{1}{a^{\frac{\beta}{\alpha}}} \right)$$

$$\leq \frac{\alpha M_2}{\beta - \alpha} \frac{1}{a^{\frac{\beta-\alpha}{\alpha}}} + \frac{M_1}{a^{\frac{\beta}{\alpha}}}.$$

These estimates together with the Theorem 7.5 of [11] imply,

$$V(f_{\alpha,\beta}; [a^{\frac{1}{\alpha}}, x)) \leq \frac{\alpha M_2}{\beta - \alpha} \frac{1}{a^{\frac{\beta-\alpha}{\alpha}}} + \frac{M_1}{a^{\frac{\beta}{\alpha}}}. \qquad (27)$$

If x tends to ∞, one gets $f_{\alpha,\beta} \in BV([a^{1/\alpha}, \infty))$. $\qquad \square$

Corollary 29 and Proposition 30 provide us Henstock-Kurzweil integrable functions defined on unbounded intervals which are not Lebesgue integrable.

Corollary 31. *Let a, α, β be such that: $0 < a$, $0 < \alpha < \beta \leq 1$ and $1 < \beta + \alpha$. Suppose that $f : [a, \infty) \to \mathbb{R}$ satisfies both the hypotheses of Corollary 29 and Proposition 30. Then, the function $f_{\alpha,\beta}$ belongs to $HK([a^{1/\alpha}, \infty)) \cap BV([a^{1/\alpha}, \infty)) \setminus L^1([a^{1/\alpha}, \infty))$.*

Taking into account the above functions we have the following corollary.

Corollary 32. *Let a, α, β be such that: $0 < a$, $0 < \alpha < \beta \leq 1$ and $1 < \beta + \alpha$, and let h in $BV([-a^{1/\alpha}, a^{1/\alpha}])$. Suppose that $f : [a, \infty) \to \mathbb{R}$ satisfies both the hypotheses of Corollary 29 and Proposition 30. Then $f : \mathbb{R} \to \mathbb{R}$ defined by*

$$g(t) = \begin{cases} h(t) & \text{if } t \in (-a^{1/\alpha}, a^{1/\alpha}), \\[2mm] \dfrac{f(|t|^\alpha)}{|t|^\beta} & \text{if } f \in (-\infty, -a^{1/\alpha}] \cup [a^{1/\alpha}, \infty) \end{cases}$$

is in $HK(\mathbb{R}) \cap BV(\mathbb{R}) \setminus L(\mathbb{R})$.

Example 33. *Let us consider the trigonometric functions $\sin(t)$ and $\cos(t)$. Then the following family of functions satisfies the hypotheses of Theorem 34.*

$$\sin_\beta^\alpha : \mathbb{R} \to \mathbb{R}; \quad \sin_\beta^\alpha(t) = \chi_{1,a}(t) \frac{\sin(t^\alpha)}{t^\beta},$$

$$\cos_\beta^\alpha : \mathbb{R} \to \mathbb{R}; \quad \cos_\beta^\alpha(t) = \chi_{2,a}(t) \frac{\cos(t^\alpha)}{t^\beta}.$$

Here $\chi_{1,\alpha}$ and $\chi_{2,\alpha}$ are the characteristic functions of the intervals $[\pi^{1/\alpha},\infty)$ and $[(\pi/2)^{1/\alpha},\infty)$, respectively. The numbers α, β are taken as in Corollary 31.

From the above example belongs to $HK(\mathbb{R}) \cap BV(\mathbb{R}) \setminus L(\mathbb{R})$. By the Multiplier theorem it follows that $HK(\mathbb{R}) \cap BV(\mathbb{R}) \subset L^2(\mathbb{R})$, so the above function is in $BV_0(\mathbb{R}) \cap L^2(\mathbb{R}) \setminus L(\mathbb{R})$. Therefore, there exist functions in $L^2(\mathbb{R}) \setminus L(\mathbb{R})$ such that their Fourier transforms exist as in (1), as an integral in HK sense.

5. The Riemann-Lebesgue Lemma and the Dirichlet-Jordan Theorem for BV_0 functions

The Riemann-Lebesgue lemma is a fundamental result of the Harmonic Analysis. An novel aspect is the validity of this lemma for functions which are not Lebesgue integrable, since this fact could help to expand the space of functions where the inversion of the Fourier transform is possible. In this section we prove a generalization of the Riemann-Lebesgue Lemma for functions of bounded variation which vanish at infinity. As consequence, it is obtained a proof of the Dirichlet-Jordan theorem for this kind of functions. This theorem provides a pointwise inversion of the Fourier transform.

We observe that the implications 1 and 2 of Theorem 26 are particularizations of the next result.

Theorem 34 (Generalization of Riemann-Lebesgue Lemma). *Let $\varphi \in HK_{loc}(\mathbb{R})$ be a function such that $\Phi(t) = \int_0^t \varphi(x)dx$ is bounded function on $\mathbb{R}$. If $f \in BV_0(\mathbb{R})$, then the function $H(w) = \int_{-\infty}^\infty f(t)\varphi(wt)dt$ is defined on $\mathbb{R} \setminus \{0\}$, it is continuous and*

$$\lim_{|w|\to\infty} H(w) = 0.$$

Proof. Given $w \in \mathbb{R}$, we define $\varphi_w(t) = \varphi(wt)$. Because of $\varphi \in HK_{loc}(\mathbb{R})$, then φ and φ_w are in $HK([0,b])$, for $b > 0$. By Jordan decomposition, there exist functions f_1 and f_2 which are nondecreasing functions belonging to $BV_0(\mathbb{R})$ such that $f = f_1 - f_2$. Hence, by Chartier-Dirichlet's Test, $f\varphi_w \in HK([0,\infty])$. By applying the Multiplier Theorem and supposing $w \neq 0$, it follows

$$\begin{aligned}\int_0^\infty f(t)\varphi(wt)dt &= -\int_0^\infty \frac{\Phi(wt)}{w}df(t) \\ &= -\int_0^\infty \frac{\Phi(wt)}{w}df_1(t) \\ &+ \int_0^\infty \frac{\Phi(wt)}{w}df_2(t),\end{aligned} \qquad (28)$$

where $df_i(t)$ is the Lebesgue-Sieltjes measure generated by f_i, $i = 1, 2$.

Let β a positive number and let M the upper bound of $|\Phi|$. For $w \in [\beta,\infty)$ we have that

$$\left| \frac{\Phi(wt)}{w} \right| \leq \frac{M}{\beta}. \tag{29}$$

Since $\Phi(wt)/w$ is continuous over $[\beta, \infty)$ and the measures $df_i(t)$ are finite, then by the Dominated Convergence Theorem applied to right side integrals in (28), it follows that

$$\lim_{w \to w_0} H(w) = H(w_0),$$

for each $w_0 \in [\beta, \infty)$. Since β is arbitrary, we obtain the continuity of H on $(0, \infty)$.

Moreover, by (28), we have for $w \in (0, \infty)$ that

$$\left| \int_0^\infty f(t)\varphi(wt)dt \right| \leq \frac{M}{|w|} Var(f; [0, \infty]).$$

Thus, we conclude that

$$\lim_{|w| \to \infty} \int_0^\infty f(t)\varphi(wt)dt = 0.$$

To complete the proof, we use similar arguments for the interval $(-\infty, 0]$. □

The above theorem confirms that $H \in C_0(\mathbb{R} \setminus \{0\})$, for each $f \in BV_0(\mathbb{R})$. As corollary we have the Riemann-Lebesgue Lemma.

Corollary 35. *If $f \in BV_0(\mathbb{R})$, then $\hat{f} \in C_0(\mathbb{R} \setminus \{0\})$.*

We know that if $g, h \in BV([a, \infty])$ then $gh \in BV([a, \infty])$. Employing this fact and Theorem 34 we get the following corollary.

Corollary 36. *Suppose that $\delta, \alpha > 0$ and $f \in BV(\mathbb{R})$, then*

$$\lim_{M \to \infty} \int_\delta^\infty \frac{f(t)}{t^\alpha} e^{-iMt} dt = 0.$$

The Sine Integral is defined as

$$Si(x) = \frac{2}{\pi} \int_0^x \frac{\sin t}{t} dt,$$

which has the properties:

1. $Si(0) = 0$, $\lim_{x \to \infty} Si(x) = 1$ and
2. $Si(x) \leq Si(\pi)$ for all $x \in [0, \infty]$.

We use the Sine Integral function in the proof of the following lemma.

Lemma 37. *Let $\delta > 0$. If $f \in BV_0(\mathbb{R})$, then*

$$\lim_{\varepsilon \to 0} \int_\delta^\infty f(t) \frac{\sin \varepsilon t}{t} dt = 0.$$

Proof. By Lemma 10 we have

$$\left| \int_\delta^\infty \frac{\sin \varepsilon t}{t} f(t) dt \right| \leq \left| \int_\delta^\infty \left(\int_{\delta\varepsilon}^{t\varepsilon} \frac{\sin u}{u} du \right) df(t) \right|. \tag{30}$$

Since for each $t \in [a, \infty)$: $\lim_{\varepsilon \to 0} \int_{\delta\varepsilon}^{t\varepsilon} \frac{\sin u}{u} du = 0$ and $\left| \int_{\delta\varepsilon}^{t\varepsilon} \frac{\sin u}{u} du \right| \leq \pi Si(\pi)$ for all $\varepsilon > 0$. Then, we obtain the result applying the Lebesgue Dominated Convergence theorem to the integral on the right in (30). $\qquad\square$

Lemma 38. *Suppose that $0 < \alpha < \beta$ or $\alpha < \beta < 0$. If $f \in BV_0(\mathbb{R})$, then for all $s \in [\alpha, \beta]$ we have*

$$\lim_{\substack{a \to -\infty \\ b \to \infty}} \int_\alpha^\beta e^{ixs} \int_a^b f(t) e^{-ist} dt ds = \int_\alpha^\beta e^{ixs} \int_{-\infty}^\infty f(t) e^{-ist} dt ds. \tag{31}$$

Proof. We will do the proof for $0 < \alpha < \beta$. Let $\widehat{f_{0b}}(s) = \int_0^b f(t) e^{-ist} dt$ and $\widehat{f_0}(s) = \int_0^\infty f(t) e^{-ist} dt$, which are continuous on $\mathbb{R} \setminus \{0\}$. Therefore the integrals in (31) exist. We know that there is $R > 0$ such that $|f(t)| \leq R$ for all $t \in \mathbb{R}$, and that for any $b > 0$: $V(f; [0, b]) \leq V(f; [0, \infty))$. For each $s \in [\alpha, \beta]$ the Multiplier theorem implies

$$\left| \widehat{f_{0b}}(s) \right| \leq \frac{2}{\alpha} \{R + V(f; [0, \infty))\} = N.$$

This inequality implies that for any $b > 0$ and all $s \in [\alpha, \beta]$: $\left| e^{ixs} \widehat{f_{0b}}(s) \right| \leq N$, for each $x \in \mathbb{R}$. Applying the theorem of Hake we have: $\lim_{b \to \infty} \widehat{f_{0b}}(s) = \widehat{f_0}(s)$. Then, by the Lebesgue Dominated Convergence theorem,

$$\lim_{b \to \infty} \int_\alpha^\beta e^{ixs} \widehat{f_{0b}}(s) ds = \int_\alpha^\beta e^{ixs} \widehat{f_0}(s) ds.$$

To conclude the proof, we follow a similar process over the interval $[a, 0]$ leading a to minus infinity. $\qquad\square$

To obtain the Dirichlet-Jordan theorem we state the following lemma [22, Theorem 11.8].

Lemma 39. *Let $\delta > 0$. If g is of bounded variation on $[0, \delta]$, then*

$$\lim_{M \to \infty} \frac{2}{\pi} \int_0^\delta g(t) \frac{\sin Mt}{t} dt = g(0+)$$

Theorem 40 (Dirichlet-Jordan Theorem). *If f is a function in $BV_0(\mathbb{R})$, then, for each $x \in \mathbb{R}$,*

$$\lim_{\substack{M \to \infty \\ \varepsilon \to 0}} \frac{1}{2\pi} \int_{\varepsilon < |s| < M} e^{ixs} \widehat{f}(s) ds = \frac{1}{2} \{ f(x+0) + f(x-0) \}. \tag{32}$$

Proof. Let $g(x, t) = f(x - t) + f(x + t)$ and suppose that $\delta > 0$. By Fubini's theorem for the Lebesgue integral [22, Theorem 15.7] at $[-M, -\varepsilon] \times [a, b]$ and $[\varepsilon, M] \times [a, b]$ and Lema 38, we have

$$\int_{\varepsilon < |s| < M} e^{ixs} \int_{-\infty}^{\infty} f(t) e^{-ist} dt ds = \lim_{\substack{a \to -\infty \\ b \to \infty}} \left(\int_{-M}^{-\varepsilon} + \int_{\varepsilon}^{M} \right) e^{ixs} \int_a^b f(t) e^{-ist} dt ds$$

$$= \lim_{\substack{a \to -\infty \\ b \to \infty}} \int_a^b f(t) \left(\int_{-M}^{-\varepsilon} + \int_{\varepsilon}^{M} \right) e^{is(x-t)} ds dt$$

$$= \int_{-\infty}^{\infty} f(t) \left(\int_{-M}^{-\varepsilon} + \int_{\varepsilon}^{M} \right) e^{is(x-t)} ds dt$$

$$= 2 \int_0^{\infty} \frac{g(x, t)}{t} (\sin M t - \sin \varepsilon t) dt$$

$$= 2 \int_{\delta}^{\infty} \frac{g(x, t)}{t} (\sin M t - \sin \varepsilon t) dt$$

$$+ 2 \int_0^{\delta} \frac{g(x, t)}{t} (\sin M t - \sin \varepsilon t) dt.$$

In $[\delta, \infty]$, by Corollary 36 and Lemma 37, we get

$$\lim_{M \to \infty, \varepsilon \to 0} \int_{\delta}^{\infty} \frac{g(x, t)}{t} (\sin M t - \sin \varepsilon t) dt = 0. \tag{33}$$

In $[0, \delta]$, applying the Lebesgue Dominate Convergence theorem,

$$\lim_{\varepsilon \to 0} \int_0^{\delta} \frac{g(x, t)}{t} \sin \varepsilon t dt = 0. \tag{34}$$

Now, by Lemma 39,

$$\lim_{M\to\infty} \int_0^\delta g(x,t)\frac{\sin Mt}{t}dt = g(x,0+) = \frac{\pi}{2}\left[f(x-0) + f(x+0)\right].$$

We conclude the proof combining (33), (34) and the above expression.   □

We observe that the classical theorem of Dirichlet-Jordan on $L(\mathbb{R})$ is a particular case of Theorem 40. Taking into account that $HK(\mathbb{R}) \cap BV(\mathbb{R}) \subset BV_0(\mathbb{R})$, then from Theorem 34 and Theorem 40 we get that if $f \in HK(\mathbb{R}) \cap BV(\mathbb{R})$, then its Fourier transform $\widehat{f}(s)$ exists in each $s \in \mathbb{R}$; $\widehat{f} \in C_0(\mathbb{R}\setminus\{0\})$, and the expression (32) holds for each $x \in \mathbb{R}$.

Corollary 41. *There exist functions in* $L^2(\mathbb{R}) \setminus L(\mathbb{R})$ *such that their Fourier transforms exist as in* (1) *and, for each* $x \in \mathbb{R}$, *the expression* (32) *is true.*

6. Conclusions

We present theorems about convergence of integrals of products in the completion of $HK(I)$, which those we have a version of Riemann-Lebesgue Lemma (over compact intervals) and analogous results at Riemann-Lebesgue property, a characterization of behavior of n-th partial sum of the Fourier series. Moreover, we have gotten basic properties (existence as integral, continuity, asymptotic behavior) about Fourier transform using Henstock-Kurzweil Integral, for this was necessary to get a generalization of Riemann-Lebesgue Lemma over $BV_0(\mathbb{R})$, in particular those characteristics are valid over $HK(\mathbb{R}) \cap BV(\mathbb{R})$. This intersection does not have relation inclusion with Lebesgue integrable functions space, we give a set of functions such that it belongs to $HK(\mathbb{R}) \cap BV(\mathbb{R}) \setminus L(\mathbb{R})$. Finally we have a generalization of Dirichlet-Jordan over $BV_0(\mathbb{R})$.

Acknowledgments

The authors acknowledge the support provided by the VIEP-BUAP, Puebla, Mexico.

Author details

Francisco J. Mendoza-Torres[1], Ma. Guadalupe Morales-Macías[2], Salvador Sánchez-Perales[3], and Juan Alberto Escamilla-Reyna[1]

1 Facultad de Ciencias Físico-Matemáticas, Benemérita Universidad Autónoma de Puebla, Puebla, Mexico

2 Departamento de Matemáticas, Universidad Autónoma Metropolitana-Iztapalapa, México, D. F., Mexico

3 Instituto de Física y Matemáticas, Universidad Tecnológica de la Mixteca, Oaxaca, Mexico

References

[1] Bachman G, Narici L, Beckenstein E. Fourier and Wavelet Analysis. New York: Springer-Verlag; 1991.

[2] Kahane Ch. Generalizations of the Riemann-Lebesgue and Cantor-Lebesgue lemmas. Czech. Math. J. 1980;30(1):108-117. Available from: http://tinyurl.com/myrs3wp

[3] Mendoza-Torres FJ, Morales-Macías MG, Escamilla-Reyna JA, Arredondo-Ruiz JH. Several aspects around the Riemann-Lebesgue Lemma. J. Adv. Res. Pure Math. 2012 Nov; 5(3), 33-46.

[4] Mendoza-Torres FJ, On pointwise inversion of the Fourier transform of BV_0 functions, Ann. Funct. Anal. 2010 Dic; 1 2010(2); 112-120. Available from: http://tinyurl.com/kgtq3vd

[5] Talvila E. Henstock-Kurzweil Fourier transforms. Illinois J. Math. 2002; 46: 1207-1226. Available from: http://tinyurl.com/kbf2llg

[6] Mendoza-Torres FJ, Escamilla Reyna JA, Sánchez-Perales S. Some results about the Henstock-Kurzweil Fourier transform. Math. Bohem. 2009; 134(4): 379–386. Available from: http://tinyurl.com/pwtsekm

[7] Zygmund A. Trigonometric Series: Vols. 1 and 2. London-New York: Cambridge University Press; 1968.

[8] Mohanty P, Talvila E. A product convergence theorem for Henstock-Kurzweil integral. Real Anal. Exch. 2003-2004;29(1): 199-204. Available from: http://tinyurl.com/nb5c7a6

[9] E.C. Titchmarsh. The order of magnitude of the coefficients in a generalized Fourier series. Proc. Lond. Math. Soc. (2) 1923-1924; 22: xxv-xxvi, doi:10.1112/plms/s2-22.1.1-s

[10] Sánchéz-Perales S. et al. Henstock-Kurzweil integral transforms. Int. J. Math. Math. Sci. 2012(2012), Article ID 209462, 11 pages. Available from: http://www.hindawi.com/journals/ijmms/2012/209462/ref/

[11] Bartle R. A Modern Theory of Integration. Graduate Studies in Mathematics, Vol. 32. Providence: American Mathematical Society; 2001.

[12] Bongiorno B., Panchapagesan TV. On the Alexiewicz topology of the Denjoy space. Real Anal. Exch. 1995-1996;21(2): 604–614. Avialable from: http://tinyurl.com/m2rbdr4

[13] Talvila E. The Distributional Denjoy integral. Real Anal. Exch. 2007;33(1): 51-84. Available from: http://tinyurl.com/m4zjn9b

[14] Kadelburg Z, Marjanović M. Interchanging two limits. Enseign. Math. 2005(8); 15-29. Availabre from: http://tinyurl.com/kyjy6p2

[15] Rudin W. Principles of Mathematical Analysis. United States of America: McGraw Hill; 1964

[16] Heuser H. Functional Analysis. New York: John Wiley and Sons; 1982.

[17] Riesz M, Livingston AE. A Short Proof of a Classical Theorem in the Theory of Fourier Integrals. Amer. Math. Monthly 62(6) Jun. - Jul. 1955, 434-437. Available from: http://www.jstor.org/stable/2307003?seq=1

[18] Talvila E. Necessary and sufficient conditions for differentiating under the integral sign. Amer. Math. Monthly 2001(108): 544-548. Available from: http://arxiv.org/pdf/math/0101012.pdf

[19] Peng-Yee L. Lanzhou Lectures on Henstock Integration. Series in Real Analysis Vol.2. Singapore: World Scientific; 1989.

[20] Swartz Ch. Introduction to Gauge Integrals. Singapore: World Scientific; 2001.

[21] Talvila E. Limits and Henstock integrals of products. Real Anal. Exch. 1999/00;25(2): 907-918. Available from: http://tinyurl.com/ke4gfkh

[22] Apostol T. Mathematical Analysis. Massachusetts: Addison-Wesley Publishing Company Reading; 1974.

2

Double Infinitesimal Fourier Transform

Takashi Gyoshin Nitta

Additional information is available at the end of the chapter

1. Introduction

For Fourier transform theory, one of the most important and difficult things is how to treat the Dirac delta function and how to define it. In 1930, the Dirac delta function was defined originally by Paul A.M. Dirac([1]) in order to create the quantum mechanical theory in Physics. In classical mechanics, there is the beautiful Newton's law. Under it, it is assumed that a particle is a point with a mass. For the investigation of a small world for example, elementary particles, it should be changed to the quantum mechanical theory where particles are not already only points as in Mathematics but also some area with infinitesimal length for us. They have some properties like waves. The Dirac delta function is defined to be realized the image of the particle in the small world. The particle changed to be the moving wave, and it becomes a set of such waves. It is called field in Physics and we need the second quantization . The quantum mechanics is developed to the quantum field theory where the delta function is much complicated to treat in the standard mathematical theory.

The delta function is usually defined as the delta measure in the functional analysis. Under the basic definition, the functional analysis is developed in the functional space for example Banach space, Hilbert space. These theory is applied to the existence problem of solutions for the ordinary and partial differential equations. On the other hand, the delta function is also defined just as a function in the extension of the real number field ([3],[4],[5],[18]) in 1962. The idea is that firstly the real and complex number fields are extended to the nonstandard universe, secondly the delta function is defined as a function in the extended universe (cf. [3]).

In this chapter the real number field and complex number field are extended twice and a higher degree of delta function is defined as a function on the space of functions. By using the secondly extended delta function, the Fourier transform theory is considered, that is called " double infinitesimal Fourier transform ". In the theory, the Poisson summation formula is also satisfied, and some important examples are calculated. The Fourier transforms of δ, δ^2, ... , and $\sqrt{\delta}$, ... can be calculated, which are constant functions as 1, infinite, ... , and infinitesimal, ...

Then the Fourier transform of the gaussian functional is also calculated. The gaussian functional means that the standard part of the image for $\alpha \in L^2$ is $\exp\left(-\pi\xi \int_{-\infty}^{\infty} \alpha^2(t)dt\right)$, for $\xi \in \mathbf{C}$ with $\mathrm{Re}(\xi) > 0$. The double infinitesimal Fourier transform is calculated as $C_\xi \exp\left(-\pi\xi^{-1} \int_{-\infty}^{\infty} \alpha^2(t)dt\right)$ for $\alpha \in L^2(\mathbf{R})$, in which C_ξ is a constant independent of α.

Finally a sort of functional is constructed in the theory that associates to Riemann's zeta function. The path integral is defined for the application in the theory, and it is shown that the path integral of the functional Z_s corresponds to Riemann's zeta function in the case that $\mathrm{Re}(s) > 1$. By using the Poisson summation formula for the functional, a relationship appears between the functional and Riemann's zeta function.

2. Infinitesimal Fourier transform

The usual universe is extended, in order to treat many stages of delta functions and functions on the space of functions. For the extension, there exists two methods, one is the second extension of the universe in the nonstandard analysis ([8],[9]) and the other is the Relative set theory in the axiomatic set theory ([13]). The first one ([8],[9]) is explained here, by using an ultrafilter .

2.1. First extension of the universe

Let Λ be an infinite set. Let F be a nonprincipal ultrafilter on Λ. For each $\lambda \in \Lambda$, let S_λ be a set. An equivalence relation $\sim$ is induced from F on $\prod_{\lambda \in \Lambda} S_\lambda$. For $\alpha = (\alpha_\lambda)$, $\beta = (\beta_\lambda)$ ($\lambda \in \Lambda$),

$$\alpha \sim \beta \iff \{\lambda \in \Lambda \,|\, \alpha_\lambda = \beta_\lambda\} \in F. \tag{1}$$

The set of equivalence classes is called *ultraproduct* of S_λ for F with respect to $\sim$. If $S_\lambda = S$ for $\lambda \in \Lambda$, then it is called *ultraproduct* of S for F and it is written as $^*S$. The set S is naturally embedded in $^*S$ by the following mapping :

$$s \ (\in S) \mapsto [(s_\lambda = s), \lambda \in \Lambda] \ (\in {}^*S) \tag{2}$$

where [ ] denotes the equivalence class with respect to the ultrafilter F. The mapping is written as $*$, and call it naturally elementary embedding. From now on, we identify the image $*(S)$ as S.

Definition 2.1.1.

Let H ($\in {}^*\mathbf{Z}$) be an infinite even number. The infinite number H is even, when for $H = [(H_\lambda), \lambda \in \Lambda]$, $\{\lambda \in \Lambda \,|\, H_\lambda$ is even$\} \in F$. The number $\frac{1}{H}$ is written as ε. We define an infinitesimal lattice space $\mathbf{L}$, an infinitesimal lattice subspace L and a space of functions $R(L)$ on L as follows :

$$\mathbf{L} := \varepsilon \, {}^*\mathbf{Z} = \{\varepsilon z \,|\, z \in {}^*\mathbf{Z}\},$$

$$L := \left\{\varepsilon z \;\middle|\; z \in {}^*\mathbf{Z}, \; -\frac{H}{2} \leq \varepsilon z < \frac{H}{2}\right\} (\subset \mathbf{L}),$$

$R(L) := \{\varphi \mid \varphi \text{ is an internal function from } L \text{ to } {}^*\mathbf{C}\}$.

The space $R(L)$ is extended to the space of periodic functions on $\mathbf{L}$ with period H. We write the same notation $R(L)$ for the space of periodic functions.

Gaishi Takeuchi([18]) introduced an infinitesimal δ function. Furthermore Moto-o Kinoshita ([4],[5]) constructed an infinitesimal Fourier transformation theory on $R(L)$. It is explained briefly.

Definition 2.1.2.

For φ, $\psi \in R(L)$, the infinitesimal δ function, the infinitesimal Fourier transformation $F\varphi$ ($\in$ $R(L)$), the inverse infinitesimal Fourier transformation $\overline{F}\varphi$ ($\in R(L)$) and the convolution $\varphi * \psi$ ($\in R(L)$) are defined as follows :

$$\delta \in R(L), \quad \delta(x) := \begin{cases} H & (x = 0) \\ 0 & (x \neq 0) \end{cases} \tag{3}$$

$$(F\varphi)(p) := \sum_{x \in L} \varepsilon \exp\left(-2\pi i p x\right) \varphi(x) \tag{4}$$

$$(\overline{F}\varphi)(p) := \sum_{x \in L} \varepsilon \exp\left(2\pi i p x\right) \varphi(x) \tag{5}$$

$$(\varphi * \psi)(x) := \sum_{y \in L} \varepsilon \varphi(x - y)\psi(y). \tag{6}$$

The definition implies the following theorem as same as the Fourier transform for the finite discrete abelian group.

Theorem 2.1.3.

(1) $\delta = F1 = \overline{F}1$,   (2) F is unitary, $F^4 = 1, \overline{F}F = F\overline{F} = 1$,

(3) $f * \delta = \delta * f = f$,   (4) $f * g = g * f$,

(5) $F(f * g) = (Ff)(Fg)$,   (6) $\overline{F}(f * g) = (\overline{F}f)(\overline{F}g)$,

(7) $F(fg) = (Ff) * (Fg)$,   (8) $\overline{F}(fg) = (\overline{F}f) * (\overline{F}g)$.

The definition implies the following proposition by the simple calculation:

Proposition 2.1.4.

If $l \in \mathbf{R}$, then

$$F\delta^l = (H)^{(l-1)}. \tag{7}$$

Examples of the infinitesimal Fourier transform for functions

The infinitesimal Fourier transforms of the gaussian function φ_ξ, $\varphi_{im} \in R(L)$ are calculated as follows: $\varphi_\xi(x) = \exp(-\xi\pi x^2)$, where $\xi \in \mathbf{C}$, $\mathrm{Re}(\xi) > 0$,

$\varphi_{im}(x) = \exp(-im\pi x^2)$, where $m \in \mathbf{Z}$.

For φ_ξ, we obtain :

Proposition 2.1.5.

$(F\varphi_\xi)(p) = c_\xi(p)\varphi_\xi(\frac{p}{\xi})$, where $c_\xi(p) = \sum_{x \in L} \varepsilon \exp(-\xi\pi(x + \frac{i}{\xi}p)^2)$ and p is an element of the lattice L.

If p is finite, then $\mathrm{st}(c_\xi(p)) = \frac{1}{\sqrt{\xi}}$.

Proof. The infinitesimal Fourier transforms of φ_ξ is :

$$(F\varphi_\xi)(p) = \sum_{x \in L} \varepsilon \exp(-2\pi ipx)\exp(-\xi\pi x^2)$$

$$= (\sum_{x \in L} \varepsilon \exp(-\xi\pi(x + \frac{i}{\xi}p)^2))\exp(-\pi\frac{1}{\xi}p^2) = c_\xi(p)\varphi_\xi(\frac{p}{\xi}) \tag{8}$$

where $c_\xi(p) = \sum_{x \in L} \varepsilon \exp(-\xi\pi(x + \frac{i}{\xi}p)^2)$. If p is finite, then $\mathrm{st}(c_\xi(p))$
$= \int_{-\infty}^{\infty} \exp\left(-\xi\pi\left(t + \frac{i}{\xi}\mathrm{st}(p)\right)^2\right) dt = \frac{1}{\sqrt{\xi}}$.

Theorem 2.1.3 (8) implies the following for c_ξ :

Proposition 2.1.6.

$$\varphi_\xi(x') = \left(\overline{F}c_\xi(p) * \left(c_{\frac{1}{\xi}}(-x)\varphi_\xi(x)\right)\right)(x'). \tag{9}$$

Proof. It is obtained : $(F\varphi_\xi)(p) = c_\xi(p)\varphi_\xi(\frac{p}{\xi})$, and put $\overline{F}$ to the above :

$(\overline{F}(F\varphi_\xi))(x) = (\overline{F}(c_\xi(p)\varphi_\xi(\frac{p}{\xi})))(x)$

$= (\overline{F}c_\xi(p) * \overline{F}\varphi_\xi(\frac{p}{\xi}))(x)$, that is, $\varphi_\xi(x) = (\overline{F}c_\xi(p) * \overline{F}\varphi_\xi(\frac{p}{\xi}))(x)$.

Now $(\overline{F}\varphi_\xi(\frac{p}{\xi}))(x) = \sum_{p \in L} \varepsilon \exp(-2\pi ipx)\exp(-\xi(\frac{p}{\xi})^2\pi)$

$= \sum_{p \in L} \varepsilon \exp(-\pi\frac{1}{\xi}(p^2 - 2\pi i\xi px)) = \left(\sum_{p \in L} \varepsilon \exp(-\frac{\pi}{\xi}(p - i\xi x)^2)\right)\varphi_\xi(x)$.

By the definition : $c_\xi(p) = \sum_{x \in L} \varepsilon \exp(-\pi\xi(x + i\frac{1}{\xi}p)^2)$, the sum
$\sum_{p \in L} \varepsilon \exp(-\frac{\pi}{\xi}(p - i\xi x)^2)$ is $c_{\frac{1}{\xi}}(-x)$. Hence

$$\varphi_\xi(x') = \left(\overline{F}c_\xi(p) * \left(c_{\frac{1}{\xi}}(-x)\varphi_\xi(x)\right)\right)(x'). \tag{10}$$

For the following proposition 2.1.7, the Gauss sum is recalled (cf.[15]) : For $z \in \mathbb{N}$, the Gauss sum $\sum_{l=0}^{z-1} \exp(-i\frac{2\pi}{z}l^2)$ is equal to $\sqrt{z}\frac{1+(-i)^z}{1-i}$.

Proposition 2.1.7. If $m|2H^2$ and $m|\frac{p}{\varepsilon}$, then

$$(F\varphi_{im})(p) = c_{im}(p)\exp(i\pi\frac{1}{m}p^2) \tag{11}$$

where $c_{im}(p) = \sqrt{\frac{m}{2}}\frac{1+i^{\frac{2H^2}{m}}}{1+i}$ for positive m and $c_{im}(p) = \sqrt{\frac{-m}{2}}\frac{1+(-i)^{\frac{2H^2}{-m}}}{1-i}$ for negative m.

Proof. $(F\varphi_{im})(p) = \sum_{x\in L} \varepsilon \exp(-im\pi x^2)\exp(-2\pi ixp)$
$= c_{im}(p)\exp(i\pi\frac{1}{m}p^2)$, where $c_{im}(p) = \sum_{x\in L} \varepsilon \exp(-im\pi(x+\frac{p}{m})^2)$.

Since $m|\frac{p}{\varepsilon}$, the element $\frac{p}{m}$ is in L. It is remarked that $\exp(-i\pi mx^2) = \exp(-i\pi m(x+H)^2)$. For positive m,

$$c_{im}(p) = \sum_{x\in L} \varepsilon \exp(-im\pi x^2) = \frac{m}{2}\left(\overline{\varepsilon\sqrt{\frac{2H^2}{m}}\frac{1+(-i)^{\frac{2H^2}{m}}}{1-i}}\right) \tag{12}$$

by the above Gauss sum. Hence $c_{im}(p) = \sqrt{\frac{m}{2}}\frac{1+i^{\frac{2H^2}{m}}}{1+i}$. For negative m, the proof is as same as the above.

2.2. Second extension of the universe

To treat a *-unbounded functional f in the nonstandard analysis, we need a second nonstandardization. Let $F_2 := F$ be a nonprincipal ultrafilter on an infinite set $\Lambda_2 := \Lambda$ as above. Denote the ultraproduct of a set S with respect to F_2 by $^*S$ as above. Let F_1 be another nonprincipal ultrafilter on an infinite set Λ_1. Take the *-ultrafilter $^*F_1$ on $^*\Lambda_1$. For an internal set S in the sense of *-nonstandardization, let $^\star S$ be the *-ultraproduct of S with respect to $^*F_1$. Thus, a double ultraproduct $^\star(^*\mathbf{R})$, $^\star(^*\mathbf{Z})$, etc are defined for the set $\mathbf{R}$, $\mathbf{Z}$, etc. It is shown easily that

$$^\star(^*\mathbf{S}) = S^{\Lambda_1 \times \Lambda_2}/F_1^{F_2}, \tag{13}$$

where $F_1^{F_2}$ denotes the ultrafilter on $\Lambda_1 \times \Lambda_2$ such that for any $A \subset \Lambda_1 \times \Lambda_2$, $A \in F_1^{F_2}$ if and only if

$$\{\lambda \in \Lambda_1 \mid \{\mu \in \Lambda_2 \mid (\lambda,\mu) \in A\} \in F_2\} \in F_1. \tag{14}$$

The work is done with this double nonstandardization. The natural imbedding $^\star S$ of an internal element S which is not considered as a set in *-nonstandardization is often denoted simply by S.

An infinite number in $^\star(^\ast\mathbf{R})$ is defined to be greater than any element in $^\ast\mathbf{R}$. We remark that an infinite number in $^\ast\mathbf{R}$ is not infinite in $^\star(^\ast\mathbf{R})$, that is, the word "an infinite number in $^\star(^\ast\mathbf{R})$" has a double meaning. An infinitesimal number in $^\star(^\ast\mathbf{R})$ is also defined to be nonzero and whose absolute value is less than each positive number in $^\ast\mathbf{R}$.

Definition 2.2.1.

Let $H(\in {}^\ast\mathbf{Z})$, $H'(\in {}^\star(^\ast\mathbf{Z}))$ be even positive numbers such that H' is larger than any element in $^\ast\mathbf{Z}$, and let $\varepsilon(\in {}^\ast\mathbf{R})$, $\varepsilon'(\in {}^\star(^\ast\mathbf{R}))$ be infinitesimals satifying $\varepsilon H = 1$, $\varepsilon'H' = 1$. We define as follows :

$$\mathbf{L} := \varepsilon\,{}^\ast\mathbf{Z} = \{\varepsilon z \mid z \in {}^\ast\mathbf{Z}\}, \quad \mathbf{L}' := \varepsilon'\,{}^\star(^\ast\mathbf{Z}) = \{\varepsilon'z' \mid z' \in {}^\star(^\ast\mathbf{Z})\},$$

$$L := \left\{ \varepsilon z \,\middle|\, z \in {}^\ast\mathbf{Z}, -\tfrac{H}{2} \le \varepsilon z < \tfrac{H}{2} \right\} (\subset \mathbf{L}),$$

$$L' := \left\{ \varepsilon'z' \,\middle|\, z' \in {}^\star(^\ast\mathbf{Z}), -\tfrac{H'}{2} \le \varepsilon'z' < \tfrac{H'}{2} \right\} (\subset \mathbf{L}').$$

Here L is an ultraproduct of lattices

$$L_\mu := \left\{ \varepsilon_\mu z_\mu \,\middle|\, z_\mu \in \mathbf{Z}, -\tfrac{H_\mu}{2} \le \varepsilon_\mu z_\mu < \tfrac{H_\mu}{2} \right\} \quad (\mu \in \Lambda_2)$$

in $\mathbf{R}$, and L' is also an ultraproduct of lattices

$$L'_\lambda := \left\{ \varepsilon'_\lambda z'_\lambda \,\middle|\, z'_\lambda \in {}^\ast\mathbf{Z}, -\tfrac{H'_\lambda}{2} \le \varepsilon'_\lambda z'_\lambda < \tfrac{H'_\lambda}{2} \right\} \quad (\lambda \in \Lambda_1)$$

in $^\ast\mathbf{R}$ that is an ultraproduct of

$$L'_{\lambda\mu} := \left\{ \varepsilon'_{\lambda\mu} z'_{\lambda\mu} \,\middle|\, z'_{\lambda\mu} \in \mathbf{Z}, -\tfrac{H'_{\lambda\mu}}{2} \le \varepsilon'_{\lambda\mu} z'_{\lambda\mu} < \tfrac{H'_{\lambda\mu}}{2} \right\} \quad (\mu \in \Lambda_2).$$

A latticed space of functions X is defined as follows,

$$\begin{aligned} X &:= \{a \mid a \text{ is an internal function with double meanings, from} {}^\star L \text{ to } L'\} \\ &= \{[(a_\lambda), \lambda \in \Lambda_1] \mid a_\lambda \text{ is an internal function from } L \text{ to } L'_\lambda\} \end{aligned} \tag{15}$$

where $a_\lambda : L \to L'_\lambda$ is $a_\lambda = [(a_{\lambda\mu}), \mu \in \Lambda_2]$, $a_{\lambda\mu} : L_\mu \to L'_{\lambda\mu}$.

Three equivalence relations $\sim_H$, $\sim_{\star(H)}$ and $\sim_{H'}$ are defined on $\mathbf{L}$, $\star(\mathbf{L})$ and $\mathbf{L}'$:

$$x \sim_H y \iff x - y \in H\,{}^\ast\mathbf{Z}, \quad x \sim_{\star(H)} y \iff x - y \in \star(H)\,{}^\star(^\ast\mathbf{Z}),$$

$$x \sim_{H'} y \iff x - y \in H'\,{}^\star(^\ast\mathbf{Z}).$$

Then $\mathbf{L}/\sim_H$, $\star(\mathbf{L})/\sim_{\star(H)}$ and $\mathbf{L}'/\sim_{H'}$ are identified as L, $\star(L)$ and L'. Since $\star(L)$ is identified with L, the set $\star(\mathbf{L})/\sim_{\star(H)}$ is identified with $\mathbf{L}/\sim_H$. Furthermore X is represented as the following internal set :

$$\{a \mid a \text{ is an internal function from } \star(\mathbf{L})/\sim_{\star(H)} \text{ to } \mathbf{L}'/\sim_{H'}\}. \tag{16}$$

The same notation is used as a function from $^\star(L)$ to L' to represent a function in the above internal set. The space A of functionals is defined as follows:

$$A := \{f \mid f \text{ is an internal function with a double meaning from } X \text{ to } ^\star(^\star \mathbf{C})\}. \qquad (17)$$

An infinitesimal delta function $\delta(a)(\in A)$, an infinitesimal Fourier transform of $f(\in A)$, an inverse infinitesimal Fourier transform of f and a convolution of f, $g(\in A)$, are defined by the following :

Definition 2.2.2. The delta function

$$\delta(a) := \begin{cases} (H')^{(^\star H)^2} & (a = 0) \\ 0 & (a \neq 0) \end{cases} \qquad (18)$$

and, with $\varepsilon_0 := (H')^{-(^\star H)^2} \in {}^\star(^\star \mathbf{R})$,

$$(Ff)(b) := \sum_{a \in X} \varepsilon_0 \exp\left(-2\pi i \sum_{k \in L} a(k)b(k)\right) f(a) \qquad (19)$$

$$(\overline{F}f)(b) := \sum_{a \in X} \varepsilon_0 \exp\left(2\pi i \sum_{k \in L} a(k)b(k)\right) f(a) \qquad (20)$$

$$(f * g)(a) := \sum_{a' \in X} \varepsilon_0 f(a - a')g(a'). \qquad (21)$$

The inner product on A is defined as:

$$(f, g) := \sum_{b \in X} \varepsilon_0 \overline{f(b)} g(b), \qquad (22)$$

where $\overline{f(b)}$ is the complex conjugate of $f(b)$. In the section 3, Riemann's zeta function is written down as a nonstandard functional in Definition 2.2.2. In general, $\sum_{k \in L} a^2(k)$ is infinite, and it is difficult to consider the meaining of F, $\overline{F}$ in Definition 2.2.2 as standard objects. They are defined only algebraically. In order to understand Definition 2.2.2 analytically for a standard one, we change the definition briefly, to Definition 2.2.3. By replacing the definitions of L', δ, ε_0, F, $\overline{F}$ in Definition 2.2.2 as the following, another type of infinitesimal Fourier transformation is defined later. The different point is only the definition of an inner product of the space of functions X. In Definition 2.2.2 , the inner product of a, $b(\in X)$ is $\sum_{k \in L} a(k)b(k)$, and in the following definition, it is $^\star \varepsilon \sum_{k \in L} a(k)b(k)$. **Definition 2.2.3.** $L' := \left\{\varepsilon'z' \mid z' \in {}^\star(^\star \mathbf{Z}), -{}^\star H\frac{H'}{2} \leq \varepsilon'z' < {}^\star H\frac{H'}{2}\right\}$,

$$\delta(a) := \begin{cases} (^\star H)^{\frac{(^\star H)^2}{2}} H'^{(^\star H)^2} & (a = 0), \\ 0 & (a \neq 0) \end{cases} \qquad (23)$$

and, with $\varepsilon_0 := (^\star H)^{-\frac{(^\star H)^2}{2}} H' - (^\star H)^2$

$$(Ff)(b) := \sum_{a \in X} \varepsilon_0 \exp\left(-2\pi i \,^\star\varepsilon \sum_{k \in L} a(k)b(k)\right) f(a) \tag{24}$$

$$(\overline{F}f)(b) := \sum_{a \in X} \varepsilon_0 \exp\left(2\pi i \,^\star\varepsilon \sum_{k \in L} a(k)b(k)\right) f(a). \tag{25}$$

Then the lattice $L'_{\lambda\mu}$ is an abelian group for each $\lambda\mu$. The following theorem is obtained as same as the case of the discrete abelian group :

Theorem 2.2.4.

(1) $\delta = F1 = \overline{F}1$,   (2) F is unitary, $F^4 = 1, \overline{F}F = F\overline{F} = 1$,

(3) $f * \delta = \delta * f = f$,   (4) $f * g = g * f$,

(5) $F(f * g) = (Ff)(Fg)$,   (6) $\overline{F}(f * g) = (\overline{F}f)(\overline{F}g)$,

(7) $F(fg) = (Ff) * (Fg)$,   (8) $\overline{F}(fg) = (\overline{F}f) * (\overline{F}g)$.

The definition directly implies the following proposition :

Proposition 2.2.5.   If $l \in \mathbf{R}^+$, then

$$F\delta^l = (H')^{(l-1)(^\star H)^2}. \tag{26}$$

If there exists α, $\beta \in L^2(\mathbf{R})$ so that $a = \,^\star\alpha|_L$, $b = \,^\star\beta|_L$, that is, $a(k) = \star(^\star\alpha(k))$, $b(k) = \star(^\star\beta(k))$, then $\mathrm{st}(\mathrm{st}(^\star\varepsilon \sum_{k \in L} a(k)b(k))) = \int_{-\infty}^\infty \alpha(x)b(x)dx$. Definition 2.2.3 is easier understanding than Definition 2.2.2 for a standard meaning in analysis. For the reason, we consider mainly Definition 2.2.3 about several examples. However Definition 2.2.2 is also treated algebraically, as algebraically defined functions are not always L^2-functions on $\mathbf{R}$. The two types of Fourier transforms are different in a standard meaning.

Examples of the double infinitesimal Fourier transform

It is defined: an equivalence relation $\sim_{\star HH'}$ in $\mathbf{L}'$ by $x \sim_{\star HH'} y \Leftrightarrow x - y \in \,^\star HH'^\star(^\star\mathbf{Z})$. The quotient space $\mathbf{L}'/\sim_{\star HH'}$ is defined with $\mathbf{L}'$. Let

$X_{H,\star HH'} := \{a' \,|\, a'$ is an internal function with a double meaning, from $^\star\mathbf{L}/ \quad \sim_{\star(H)}$ to $\mathbf{L}'/\sim_{\star HH'}\}$

and let $\mathbf{e}$ be a mapping from X to $X_{H,\star HH'}$, defined by $(\mathbf{e}(a))([k]) = [a(\hat{k})]$, where $[\;\;]$ on the left-hand side represents the equivalence class for the equivalence relation $\sim_{\star(H)}$ in $^\star\mathbf{L}$, $\hat{k}$ is a representative in $^\star(L)$ satisfying $k \sim_{\star(H)} \hat{k}$, and $[\;\;]$ on the right-hand side represents the equivalence class for the equivalence relation $\sim_{\star HH'}$ in $\mathbf{L}'$. Furthermore $f(a')$ is identified to be $f(\mathbf{e}^{-1}(a'))$.

The double infinitesimal Fourier transform of $\exp\left(-\pi^\star \varepsilon \tilde{\zeta} \sum_{k\in L} a^2(k)\right)$

The double infinitesimal Fourier transform of

$$g_{\tilde\zeta}(a) = \exp\left(-\pi^\star \varepsilon \tilde\zeta \sum_{k\in L} a^2(k)\right), \tag{27}$$

where $\tilde\zeta \in \mathbf{C}$, $\mathrm{Re}(\tilde\zeta) > 0$,

is calculated in the space A of functionals, for Definition 2.2.3. It is identified $^\star(^\star\tilde\zeta) \in \mathbf{C}$ with $\tilde\zeta \in \mathbf{C}$.

Theorem 2.2.6. $F(g_{\tilde\zeta})(b) = C_{\tilde\zeta}(b)g_{\tilde\zeta}(\frac{b}{\tilde\zeta})$, where $b \in X$ and

$$C_{\tilde\zeta}(b) = \sum_{a\in X} \varepsilon_0 \exp\left(-\pi^\star \varepsilon \tilde\zeta \sum_{k\in L}(a(k) + i\frac{1}{\tilde\zeta}b(k))^2\right). \tag{28}$$

Proof. The infinitesimal Fourier transform of $g_{\tilde\zeta}(a)$ is done.

$F(g_{\tilde\zeta})(b) = F\left(\exp\left(-\pi^\star \varepsilon \tilde\zeta \sum_{k\in L} a^2(k)\right)\right)(b)$

$= \sum_{a\in X} \varepsilon_0 \exp\left(-2i\pi^\star \varepsilon \sum_{k\in L} a(k)b(k)\right)\exp\left(-\pi^\star \varepsilon \tilde\zeta \sum_{k\in L} a^2(k)\right)$

$= C_{\tilde\zeta}(b)g_{\tilde\zeta}(\frac{b}{\tilde\zeta})$.

Let $\star \circ \star : \mathbf{R} \to {}^\star({}^\star\mathbf{R})$ be the natural elementary embedding and let $\mathbf{st}(c)$ for $c \in {}^\star({}^\star\mathbf{R})$ be the standard part of c with respect to the natural elementary embedding $\star \circ \star$. Let $st(c)$ be the standard part of c with respect to the natural elementary embedding $\star$ and $\star$. Then $\mathbf{st} = st \circ st$.

Theorem 2.2.7. If the image of b $(\in X)$ is bounded by a finite value of $^\star\mathbf{R}$, that is, there exists $b_0 \in {}^\star\mathbf{R}$ such that $k \in L \Rightarrow |b(k)| \leq \star(b_0)$, then

$$\mathbf{st}(C_{\tilde\zeta}(b)) = \left(\star\left(\frac{1}{\sqrt{\tilde\zeta}}\right)\right)^{H^2} (\in {}^\star\mathbf{R}), \quad \mathbf{st}\left(\frac{C_{\tilde\zeta}(b)}{\star\left(\left(\star\left(\frac{1}{\sqrt{\tilde\zeta}}\right)\right)^{H^2}\right)}\right) = 1. \tag{29}$$

Proof. $\mathbf{st}(C_{\tilde\zeta}(b)) = \mathbf{st}(\sum_{a\in X}\prod_{k\in L}\sqrt{\varepsilon\varepsilon'}\exp\left(-\pi\tilde\zeta\{\sqrt{\varepsilon}(a(k)) + i\sqrt{\varepsilon}\frac{1}{\tilde\zeta}(b(k))\}^2\right))$

$= \prod_{k\in L}\int_{-{}^\star\infty}^{{}^\star\infty}\exp\left(-\pi\tilde\zeta\{x + i\sqrt{\varepsilon}\frac{1}{\tilde\zeta}st_2(b(k))\}^2\right)dx$

$= \prod_{k\in L}\int_{-{}^\star\infty}^{{}^\star\infty}\exp\left(-\pi\tilde\zeta x^2\right)dx$.

The argument is same about the infinitesimal Fourier transform of $g'_{\tilde\zeta}(a) = \exp(-\pi\tilde\zeta\sum_{k\in L}a^2(k))$, for Definition 2.2.2, as the above.

Theorem 2.2.8.

$$F(g'_\varsigma)(b) = B_\varsigma(b)g'_\varsigma(\frac{b}{\varsigma}), \tag{30}$$

where $b \in X$ and

$B_\varsigma(b) = \sum_{a \in X} \varepsilon_0 \exp\left(-\pi\varsigma \sum_{k \in L}(a(k) + i\frac{1}{\varsigma}b(k))^2\right)$. Furthermore, if the image of b ($\in X$) is bounded by a finite value of $*\mathbf{R}$, that is, $\exists b_0 \in *\mathbf{R}$ s.t. $k \in L \Rightarrow |b(k)| \leq \star(b_0)$ then

$$\mathrm{st}(B_\varsigma(b)) = \left(\ast\left(\frac{1}{\sqrt{\varsigma}}\right)\right)^{H^2} (\in *\mathbf{R}), \quad \mathrm{st}\left(\frac{B_\varsigma(b)}{\ast\left(\left(\ast\left(\frac{1}{\sqrt{\varsigma}}\right)\right)^{H^2}\right)}\right) = 1. \tag{31}$$

The double infinitesimal Fourier transform of $\exp(-i\pi m^\star \varepsilon \sum_{k \in L} a^2(k))$

The double infinitesimal Fourier transform of $g_{im}(a) = \exp(-i\pi m^\star \varepsilon \sum_{k \in L} a^2(k))$, where $m \in \mathbf{Z}$, is calculated for Definition 2.2.3.

Proposition 2.2.9. $F(g_{im})(b)$ is written as $C_{im}(b)g_{\frac{1}{im}}(b)$.

If $m|2^\star HH'^2$ and $m|\frac{b(k)}{\varepsilon'}$ for an arbitrary k in L, then $F(g_{im})(b) = C_{im}(b)g_{\frac{1}{im}}(b)$, where $C_{im}(b) = \left(\sqrt{\frac{m}{2}}\frac{1+i^{\frac{2^\star HH'^2}{m}}}{1+i}\right)^{(^\star H)^2}$ for a positive m and

$C_{im}(b) = \left(\sqrt{\frac{-m}{2}}\frac{1+(-i)^{\frac{2^\star HH'^2}{-m}}}{1-i}\right)^{(^\star H)^2}$ for a negative m.

Proof.

$F(g_{im})(b) = C_{im}(b)g_{\frac{1}{im}}(b)$, where $C_{im}(b) = \sum_{a \in X} \varepsilon_0 \exp(-i\pi m^\star \varepsilon \sum_{k \in L}(a(k) + \frac{1}{m}b(k))^2)$.

When $a(k)$, $b(k)$ are denoted as $\varepsilon'n'$, $\varepsilon'l'$,

$\sum_{-^\star H\frac{H'^2}{2} \leq a(k) < ^\star H\frac{H'^2}{2}} \exp(-i\pi m^\star \varepsilon \sum_{k \in L}(a(k) + \frac{1}{m}b(k))^2$

$$= \sum_{-^\star H\frac{H'^2}{2} \leq \varepsilon'n' < ^\star H\frac{H'^2}{2}} \exp(-i\pi m^\star \varepsilon \sum_{k \in L}(\varepsilon'n' + \varepsilon'\frac{n'}{m})^2). \tag{32}$$

Since $m|\frac{b(k)}{\varepsilon'}$, for a positive m, it is equal to

$$\sum_{-^\star H\frac{H'^2}{2} \leq \varepsilon'n' < ^\star H\frac{H'^2}{2}} \exp(-i\pi m^\star \varepsilon\varepsilon'^2 n'^2) = \frac{m}{2}\sqrt{\frac{2^\star HH'^2}{m}}\frac{1+i^{\frac{2^\star HH'^2}{m}}}{1+i} \tag{33}$$

by Proposition 2.1.5. Hence $C_{im} = \left(\sqrt{\frac{m}{2}} \frac{1+i\frac{2^{\star}HH'^2}{m}}{1+i}\right)^{(^{\star}H)^2}$ for a positive m. For a negative m, the proof is as same as the above.

The argument for the infinitesimal Fourier transform of $g'_{im}(a) = \exp(-i\pi m \sum_{k \in L} a^2(k))$, for Definition 2.2.2, is as same as the above one of g_{im} for Definition 2.2.3.

Proposition 2.2.10. If $m|2^{\star}HH'^2$ and $m|\frac{b(k)}{\varepsilon'}$ for an arbitrary k in L, then $(F((g'_{im}))(b) = B_{im}(b)g'_{\frac{1}{im}}(b)$, where $B_{im}(b) = \left(\sqrt{\frac{m}{2}}\frac{1+i\frac{2H'^2}{m}}{1+i}\right)^{(^{\star}H)^2}$ for a positive m and $B_{im}(b) = \left(\sqrt{\frac{-m}{2}}\frac{1+(-i)\frac{2H'^2}{-m}}{1-i}\right)^{(^{\star}H)^2}$ for a negative m.

2.3. The meaning of the double infinitesimal Fourier transform

There exists a natural injection from a space of standard functions to X as

$$\alpha \mapsto (a : k \in L \mapsto \star(^{\star}\alpha(k)) \in L'). \tag{34}$$

Hence a space of standard functions is embedded in X through the natural injection. If there is no confusion, standard functions are identified as nonstandard functions by the natural injection.

For a standard functional f, if the domain of $\star(^{\star}f)$ is in X, we can define a Fourier transform $F(\star(^{\star}f))$. Since $\text{st}(\text{st}(F(\star(^{\star}f))))$ is a standard functional as $\text{st}(\text{st}(F(\star(^{\star}f))))(\alpha) = \text{st}(\text{st}(F(\star(^{\star}f))(a)))$ for $a : k \in L \to \star(^{\star}\alpha(k)) \in L'$, such standard functional has a Fourier transform $\text{st}(\text{st}(F(\star(^{\star}f))))$.

Similarly to the case of functions, the following subspace $\mathcal{L}^2(A)$ of A is defined:

Definition 2.3.1.

$$\mathcal{L}^2(A) := \{f \in A| \text{ there exists } c \in {}^{\star}\mathbf{R} \text{ so that } (\frac{1}{c}\sum_{a \in X}\varepsilon_0|f(a)|^2) < +\infty\}. \tag{35}$$

The standard part $\text{st}(\sum_{a \in X}\varepsilon_0|f(a)|^2)$ is a $*-$ norm in $\mathcal{L}^2(A)$. Theorem 2.1.3 (2) implies the following proposition.

Proposition 2.3.2. The Fourier transform F and the inverse $\overline{F}$ preserve the space $\mathcal{L}^2(A)$.

Hence if f is a standard functional so that $\star(^{\star}f)$ is an element of $\mathcal{L}^2(A)$, the Fourier transformation $F(\star(^{\star}f))$, $\overline{F}(\star(^{\star}f))$ are also in $\mathcal{L}^2(A)$. Now there is no theory of Fourier transform for functionals in "standard analysis", and it is well-known that there is no nontrivial translation-invariant measure on an infinite-dimensional separable Banach space. In fact, on the infinite-dimensional Banach space there is an infinite sequence of pairwise disjoint open balls of same sizes in a larger ball. The measure is translation-invariant, the measure of the small balls are same, but the measure of the larger ball is finite, it is contradiction. By the reason we do not argue a relationship between our Fourier transform and standard Fourier transform, any more.

Here number fields are extended twice to realize the delta function for functionals. The extended real number field divided to very small infinitesimal lattices. These lattices are too small for normal real number field and the first extended real number field to observe them. Axiomatically, the double extended number field can be treat in a large universe, that is, relative set theory ([13],[14]). The concept of observable and relatively observable are formulated, and two kinds of delta functions are defined. The Fourier transform theory is developed, which is called divergence Fourier transform . It is applied to solve an elementary ordinary differential equation with a delta function(cf.[12]).

3. Poisson summation formula

The Poisson summation formula is a fundamental formula for each Fourier transform theory. In this section, it is explained about the Kinoshita's Fourier transform and our double infinitesimal Fourier transform . Some examples of the gaussian type functions are calculated for the applications of the Poisson summation formula.

3.1. Poisson summation formula for infinitesimal Fourier transform

The Poisson summation formula of finite group is extended to Kinoshita's infinitesimal Fourier transform.

Formulation

Theorem 3.1.1. Let S be an internal subgroup of L. Then the following formula is obtained, for $\varphi \in R(L)$,

$$|S^{\perp}|^{-\frac{1}{2}} \sum_{p \in S^{\perp}} (F\varphi)(p) = |S|^{-\frac{1}{2}} \sum_{x \in S} \varphi(x) \qquad (36)$$

where $S^{\perp} := \{p \in L \mid \exp(2\pi i p x) = 1 \text{ for } \forall x \in S\}$.

Since L is an internal cyclic group, the group S is also an internal cyclic group. The generator of L is ε. The generator of S is written as εs ($s \in {}^{*}\mathbf{Z}$). Since the order of L is H^2, so s is a factor of H^2.

The following lemma is prepared for the proof of Theorem 3.1.1.

Lemma 3.1.2. $S^{\perp} = < \varepsilon \frac{H^2}{s} >$.

Proof of Lemma 3.1.2. For $p \in S^{\perp}$, we write $p = \varepsilon t$. Then the following is obtained:

$$\exp(2\pi i p \varepsilon s) = 1 \iff \exp(2\pi i \varepsilon t \varepsilon s) = 1 \iff \exp(2\pi i t \frac{s}{H^2}) = 1 \iff t \frac{s}{H^2} \in {}^{*}\mathbf{Z}. \qquad (37)$$

Hence the generater of $S^{\perp}$ is $\varepsilon\frac{H^2}{s}$.

Proof of Theorem 3.1.1. By Lemma 2.1.2, $|S| = \frac{H^2}{s}$ and $|S^{\perp}| = s$. If $x \notin S$, then $\varepsilon\frac{H^2}{s}xs = \varepsilon H^2 x \in {}^*\mathbf{Z}$, and $\left(\exp\left(2\pi i\varepsilon\frac{H^2}{s}x\right)\right)^s = 1$. For $x \in L$,

$$\sum_{p \in S^{\perp}} \exp(2\pi i p x) = \begin{cases} \dfrac{\exp(2\pi i(-\frac{H}{2})x)(1-(\exp(2\pi i\varepsilon\frac{H^2}{s}x)^s))}{1-\exp(2\pi i\varepsilon\frac{H^2}{s}x)} & (x \notin S) \\ \sum_{p \in S^{\perp}} 1 & (x \in S) \end{cases}$$

$$= \begin{cases} 0 & (x \notin S) \\ s & (x \in S) \end{cases}. \tag{38}$$

Hence

$$\sum_{p \in S^{\perp}}(F\varphi)(p) = \sum_{p \in S^{\perp}} \varepsilon(\sum_{x \in L} \varphi(x)\exp(2\pi i p x))$$

$$= \varepsilon \sum_{x \in L} \varphi(x)(\sum_{p \in S^{\perp}} \exp(2\pi i p x)) = \frac{s}{H} \sum_{x \in S} \varphi(x). \tag{39}$$

Thus

$$\frac{1}{\sqrt{s}} \sum_{p \in S^{\perp}}(F\varphi)(p) = \sqrt{\frac{s}{H^2}} \sum_{x \in S} \varphi(x) \tag{40}$$

hence

$$|S^{\perp}|^{-\frac{1}{2}} \sum_{p \in S^{\perp}}(F\varphi)(p) = \frac{1}{|S|^{\frac{1}{2}}} \sum_{x \in S} \varphi(x) \cdots (\sharp_1). \tag{41}$$

Proposition 3.1.3. Especially if s is equal to H, then $(\sharp_1)$ implies that $\sum_{p \in S^{\perp}}(F\varphi)(p) = \sum_{x \in S} \varphi(x)$. The standard part of it is $\text{st}(\sum_{p \in S^{\perp}}(F\varphi)(p)) = \text{st}(\sum_{x \in S} \varphi(x))$.

If there exists a standard function $\varphi' : \mathbf{R} \to \mathbf{C}$ so that $\varphi = {}^*\varphi'|_L$, then the right hand side is equal to $\sum_{-\infty < x < \infty} \varphi'(x)$, that is, $\sum_{-\infty < x < \infty} \text{st}(\varphi)(x)$. Furthermore if εs is infinitesimal and φ' is integrable on $\mathbf{R}$, then

$\text{st}(\varepsilon s \sum_{x \in S} \varphi(x)) = \int_{-\infty}^{\infty} \varphi'(x)dx$.

Since $(\sharp_1)$ implies that

$\sum_{p \in S^{\perp}}(F\varphi)(p) = \varepsilon s \sum_{x \in S} \varphi(x)$,

it is obtained $\text{st}(\sum_{p \in S^{\perp}}(F\varphi)(p)) = \int_{-\infty}^{\infty} \varphi'(x)dx$, that is, $\int_{-\infty}^{\infty} \text{st}(\varphi)(x)dx$.

The even number H is decomposed to prime factors $H = p_1^{l_1} p_2^{l_2} \cdots p_m^{l_m}$, where $p_1 = 2$, $p_1 < p_2 < \cdots < p_m$, each p_i is a prime number, $0 < l_i$. Since S is a subgroup of L, the number s is a factor of H^2. When we write s as $p_1^{k_1} p_2^{k_2} \cdots p_m^{k_m}$, the order of S is equal to $p_1^{2l_1-k_1} p_2^{2l_2-k_2} \cdots p_m^{2l_m-k_m}$ and the order of $S^\perp$ is $p_1^{k_1} p_2^{k_2} \cdots p_m^{k_m}$. Hence (27) is

$$(p_1^{k_1} p_2^{k_2} \cdots p_m^{k_m})^{-\frac{1}{2}} \sum_{p \in S^\perp} (F\varphi)(p)) = (p_1^{2l_1-k_1} p_2^{2l_2-k_2} \cdots p_m^{2l_m-k_m})^{-\frac{1}{2}} \sum_{x \in S} \varphi(x). \tag{42}$$

Examples

Theorem 3.1.1 is applied to the following two kinds of functions :

$$1. \varphi_i(x) = \exp(-i\pi x^2) \tag{43}$$
$$2. \varphi_\xi(x) = \exp(-\xi\pi x^2) \tag{44}$$

where $\xi \in \mathbf{C}$, $\mathrm{Re}(\xi) > 0$. Then the infinitesimal Fourier transforms are :

$$1. (F\varphi_i)(p) = \exp(-i\frac{\pi}{4}) \overline{\varphi_i(p)} \cdots (\sharp_2) \tag{45}$$
$$2. (F\varphi_\xi)(p) = c_\xi(p) \varphi_\xi(\frac{p}{\xi}), \tag{46}$$

where $\mathrm{st}(c_\xi(p)) = \frac{1}{\sqrt{\xi}}$, if p is finite. Hence the following formulas are obtained :

$$1. |S^\perp|^{-\frac{1}{2}} \exp(-i\frac{\pi}{4}) \sum_{p \in S^\perp} \overline{\varphi_i(p)} = |S|^{-\frac{1}{2}} \sum_{x \in S} \varphi_i(x), \tag{47}$$
$$2. |S^\perp|^{-\frac{1}{2}} \sum_{p \in S^\perp} c_\xi(p) \varphi_\xi(\frac{p}{\xi}) = |S|^{-\frac{1}{2}} \sum_{x \in S} \varphi_\xi(x). \tag{48}$$

When the generator of S is εs, this is written as the following, explicitly :

$$1. H \exp(-i\frac{\pi}{4}) \sum_{p \in S^\perp} \exp(i\pi p^2) = s \sum_{x \in S} \exp(-i\pi x^2) \tag{49}$$
$$2. H \sum_{p \in S^\perp} c_\xi(p) \exp(-\frac{1}{\xi}\pi p^2) = s \sum_{x \in S} \exp(-\xi\pi x^2). \tag{50}$$

The following proposition is obtained:

Proposition 3.1.4.

(i) If $s = H$, then the generator of S is 1 and $S = S^\perp = L \cap {}^*\mathbf{Z}$. Hence

$$1.\exp(-i\frac{\pi}{4})\sum_{p\in L\cap {}^*Z}\exp(i\pi p^2) = \sum_{x\in L\cap {}^*Z}\exp(-i\pi x^2) \tag{51}$$

$$2.\sum_{p\in L\cap {}^*Z} c_\xi(p)\exp(-\frac{1}{\xi}\pi p^2) = \sum_{x\in L\cap {}^*Z}\exp(-\xi\pi x^2). \tag{52}$$

Taking their standard parts, we obtain :

$$2.st(\sum_{p\in L\cap {}^*Z} c_\xi(p)\exp(-\frac{1}{\xi}\pi p^2)) = st(\sum_{x\in L\cap {}^*Z}\exp(-\xi\pi x^2))$$

$$= \sum_{-\infty<n<\infty}\exp(-\xi\pi n^2) = \theta(i\xi) \tag{53}$$

where $\theta(z)$ is a θ-function, defined by $\theta(z) = \sum_{-\infty<n<\infty}\exp(i\pi zn^2)$.

(ii) If εs is infinitesimal, then the equation:

$H\sum_{p\in S^\perp} c_\xi(p)\exp(-\frac{1}{\xi}\pi p^2) = s\sum_{x\in S}\exp(-\xi\pi x^2)$ implies the following:

$$st(\sum_{p\in S^\perp} c_\xi(p)\exp(-\frac{1}{\xi}\pi p^2)) = st(\varepsilon s\sum_{x\in S}\exp(-\xi\pi x^2))$$

$$= \int_{-\infty}^{\infty}\exp(-\xi\pi x^2)dx = \frac{1}{\sqrt{\xi}}. \tag{54}$$

It is known that $st(c_\xi(p)) = \frac{1}{\sqrt{\xi}}$, and $\sum_{-\infty<x<\infty}\exp(-\xi\pi x^2)$ in the formula 2 of (i) is equal to $\frac{1}{\sqrt{\xi}}\sum_{-\infty<p<\infty}\exp(-\frac{1}{\xi}\pi p^2)$ by the standard Poisson summation formula. Hence, by 2 of (i), we obtain $st(\sum_{p\in S^\perp} c_\xi(p)\exp(-\frac{1}{\xi}\pi p^2)) = \sum_{-\infty<p<\infty}st(c_\xi(p)\exp(-\frac{1}{\xi}\pi p^2))$.

The formula ($\sharp_2$) in 1 for $\varphi_i(x)$ is extended to $\varphi_{im}(x) = \exp(-im\pi x^2)$, for an integer m so that $m|2H^2$. If $m|\frac{p}{\varepsilon}$, we recall

$$(F\varphi_{im})(p) = c_{im}(p)\exp(i\pi\frac{1}{m}p^2),$$

where $c_{im}(p) = \sqrt{\frac{m}{2}}\frac{1+i^{\frac{2H^2}{m}}}{1+i}$ for a positive m and $c_{im}(p) = \sqrt{\frac{-m}{2}}\frac{1+(-i)^{\frac{2H^2}{-m}}}{1-i}$ for a negative m.

Hence $|S^\perp|^{-\frac{1}{2}}\sum_{p\in S^\perp} c_{im}(p)\varphi_{\frac{1}{im}}(p) = |S|^{-\frac{1}{2}}\sum_{x\in S}\varphi_{im}(x)$. When the generator $\varepsilon s'$ of $S^\perp$ satifies $m|s'$, that is, the generator εs of S satifies $m|\frac{H^2}{s}$, it reduces to the following:

$$H\sqrt{\frac{m}{2}}\frac{1+i^{\frac{2H^2}{m}}}{1+i}\sum_{p\in S^\perp}\exp(i\pi\frac{1}{m}p^2) = s\sum_{x\in S}\exp(-im\pi x^2) \tag{55}$$

for a positive m,

$$H\sqrt{\frac{-m}{2}}\frac{1+(-i)^{\frac{2H^2}{-m}}}{1-i}\sum_{p\in S^\perp}\exp(i\pi\frac{1}{m}p^2) = s\sum_{x\in S}\exp(-im\pi x^2) \tag{56}$$

for a negative m.

3.2. Poisson summation formula for Definition 2.2.2

Poisson summation formula of finite group is extended to the double infinitesimal Fourier transform for Definition 2.2.2 on the space of functionals.

Formulation

Theorem 3.2.1. Let Y be an internal subgroup of X. Then the following is obtained, for $f \in A$,

$$|Y^\perp|^{-\frac{1}{2}}\sum_{b\in Y^\perp}(Ff)(b) = |Y|^{-\frac{1}{2}}\sum_{a\in Y}f(a) \tag{57}$$

where $Y^\perp := \{b \in X \mid \exp(2\pi i < a,b >) = 1$ for $\forall a \in X\}$ and $< a,b >:= \sum_{k\in L}a(k)b(k)$.

Lemma 3.2.2. $|Y^\perp| = \frac{|X|}{|Y|}$.

Proof of Lemma 3.2.2. For $k \in L$, we denote $Y_k := \{a(k) \in L' \mid a \in Y\}$.

$b \in Y^\perp \iff \forall a \in Y$, $\exp(2\pi i \sum_{k\in L}a(k)b(k)) = 1$

$\iff \forall k \in L$, $b(k) \in Y_k^\perp$

$\iff b : L \to L'$, $\forall k \in L$, $b(k) \in Y_k^\perp$.

Hence $|Y^\perp| = \prod_{k\in L}|Y_k^\perp|$. Lemma 3.1.2 implies $|Y_k^\perp| = \frac{H'^2}{|Y_k|}$. Thus

$$|Y^\perp| = \prod_{k\in L}\left(\frac{H'^2}{|Y_k|}\right) = \frac{H'^{2*H^2}}{\prod_{k\in L}|Y_k|} = \frac{|X|}{|Y|} \tag{58}$$

Proof of Theorem 3.2.1.

$$|Y^\perp|^{-\frac{1}{2}}\sum_{b\in Y^\perp}(Ff)(b) = |Y^\perp|^{-\frac{1}{2}}\sum_{a\in X}\varepsilon_0(\sum_{b\in Y^\perp}\exp(-2\pi i < a,b >))f(a). \tag{59}$$

Since $\sum_{b\in Y^\perp}\exp(-2\pi i <a,b>) = \begin{cases} 0 & (a\notin Y) \\ |Y^\perp| & (a\in Y) \end{cases}$, the above is equal to

$$|Y^\perp|^{-\frac{1}{2}}\varepsilon_0|Y^\perp|\sum_{a\in Y}f(a) = |Y^\perp|^{\frac{1}{2}}H'^{-*H^2}\sum_{a\in Y}f(a) = |Y|^{-\frac{1}{2}}\sum_{a\in Y}f(a). \tag{60}$$

In the special case where $f(a) = \prod_{k\in L}f_k(a(k))$,

$(Ff)(b) = \sum_{a\in X}\varepsilon_0\exp(-2\pi i\sum_{k\in L}a(k)b(k))\prod_{k\in L}f_k(a(k))$

$$= \prod_{k\in L}(\sum_{a(k)\in L'}\varepsilon'\exp(-2\pi i a(k)b(k))f_k(a(k))). \tag{61}$$

Namely, the Fourier transform in functional space is the product of those in function space.

Corollary 3.2.3.

(i) If each generator of Y_k is equal to 1, f is written as $\prod_{k\in L}f_k$, $f_k = {}^*(\mathrm{st}(f_k))|_{L'}$, and $\sum_{-\infty<n<\infty}\mathrm{st}(f_k)(n)$ converges, then

$$\mathrm{st}(\sum_{b\in Y^\perp}(Ff)(b)) = \prod_{k\in L}(\sum_{-\infty<n<\infty}\mathrm{st}(f_k)(n)). \tag{62}$$

(ii) If each generator of Y_k is infinitesimal, f is written as $\prod_{k\in L}f_k$, $f_k = {}^*(\mathrm{st}(f_k))|_{L'}$ and $\mathrm{st}(f_k)$ is L_1-integrable on $\mathbf{R}$, then

$$\mathrm{st}(\sum_{b\in Y^\perp}(Ff)(b)) = \prod_{k\in L}\int_{-\infty<t<\infty}\mathrm{st}(f_k)(t)dt. \tag{63}$$

Examples

Theorem 3.2.1 is applied to the following two kinds of functionals :

$$1. f_i(a) = \exp(-i\pi\sum_{k\in L}a(k)^2) \tag{64}$$

$$2. f_\xi(a) = \exp(-\xi\pi\sum_{k\in L}a(k)^2), \tag{65}$$

where $\xi\in\mathbf{C}$, $\mathrm{Re}(\xi) > 0$.

The infinitesimal Fourier transforms of the functionals are :

$$1.(Ff_i)(b) = (-1)^{\frac{H}{2}} \overline{f_i(b)} \cdots (\sharp 3) \tag{66}$$

$$2.(Ff_{\xi})(b) = B_{\xi}(b)f_{\xi}(\frac{b}{\xi}), \tag{67}$$

hence the followings are obtained :

$$1.|Y^{\perp}|^{-\frac{1}{2}}(-1)^{\frac{H}{2}} \sum_{b \in Y^{\perp}} \overline{f_i(b)} = |Y|^{-\frac{1}{2}} \sum_{a \in Y} f_i(a) \tag{68}$$

$$2.|Y^{\perp}|^{-\frac{1}{2}} \sum_{b \in Y^{\perp}} B_{\xi}(b)f_{\xi}(\frac{b}{\xi}) = |Y|^{-\frac{1}{2}} \sum_{a \in Y} f_{\xi}(a). \tag{69}$$

These are written as the following, explicitly :

$$1.|Y^{\perp}|^{-\frac{1}{2}}(-1)^{\frac{H}{2}} \sum_{b \in Y^{\perp}} \exp(-i\pi \sum_{k \in L} b(k)^2) = |Y|^{-\frac{1}{2}} \sum_{a \in Y} \exp(-i\pi \sum_{k \in L} a(k)^2), \tag{70}$$

$$2.|Y^{\perp}|^{-\frac{1}{2}} \sum_{b \in Y^{\perp}} B_{\xi}(b) \exp(-\frac{1}{\xi}\pi \sum_{k \in L} b(k)^2) = |Y|^{-\frac{1}{2}} \sum_{a \in Y} \exp(-\xi\pi \sum_{k \in L} a(k)^2). \tag{71}$$

Corollary 3.2.3 implies the following proposition 3.2.4.

Proposition 3.2.4.

(i) If each generator of Y_k is equal to 1, then

$$1.(-1)^{\frac{H}{2}} st(\sum_{b \in Y^{\perp}} \exp(-i\pi \prod_{k \in L} b(k)^2)) = (\sum_{-\infty < n < \infty} \exp(-i\pi n^2))^{H^2} \tag{72}$$

$$2.st(\sum_{b \in Y^{\perp}} B_{\xi}(b) \exp(-\frac{1}{\xi}\pi \sum_{k \in L} b(k)^2)) = (\sum_{-\infty < n < \infty} \exp(-\xi\pi n^2))^{H^2} \tag{73}$$

$$\left(= (\theta(i\xi))^{H^2} \right).$$

(ii) If each generator of Y_k is equal to a natural number m_k, then

$$1. (-1)^{\frac{H}{2}} st\left(\sum_{b \in Y^\perp} \exp\left(-i\pi \prod_{k \in L} b(k)^2\right)\right) = \prod_{k \in L}\left(m_k \sum_{-\infty < n < \infty} \exp(-i\pi m_k^2 n^2)\right) \quad (74)$$

$$2. st\left(\sum_{b \in Y^\perp} B_{\xi}(b) \exp\left(-\frac{1}{\xi}\pi \sum_{k \in L} b(k)^2\right)\right) = \prod_{k \in L}\left(m_k \sum_{-\infty < n < \infty} \exp(-\xi\pi m_k^2 n^2)\right) \quad (75)$$

$$\left(= \prod_{k \in L}(m_k \theta(im_k^2 \xi)) \right).$$

(iii) If each generator of Y_k is infinitesimal, then

$$2. st\left(\sum_{b \in Y^\perp} B_{\xi}(b) \exp\left(-\frac{1}{\xi}\pi \sum_{k \in L} b(k)^2\right)\right) = \left(\int_{-\infty}^{\infty} \exp(-\xi\pi t^2)dt\right)^{H^2} \quad (76)$$

$$\left(= \left(*\left(\frac{1}{\sqrt{\xi}}\right)\right)^{H^2}\right).$$

The above formula ($\sharp 3$) for $f_i(a)$ is extended to $f_{im}(a) = \exp(-im\pi \sum_{k \in L} a^2(k))$, for an integer m so that $m | 2H'^2$. If $m | \frac{b(k)}{\varepsilon'}$, we recall

$$(F f_{im})(b) = B_{im}(b)f_{\frac{1}{im}}(b) \quad (77)$$

where $B_{im}(b) = \left(\sqrt{\frac{m}{2}}\frac{1+i^{\frac{2H'^2}{m}}}{1+i}\right)^{(*H)^2}$ for a positive m , $B_{im}(b) = \left(\sqrt{\frac{-m}{2}}\frac{1+(-i)^{\frac{2H'^2}{-m}}}{1-i}\right)^{(*H)^2}$ for a negative m.

Hence $|Y^\perp|^{-\frac{1}{2}} \sum_{b \in Y^\perp} B_{im}(b)f_{\frac{1}{im}}(b) = |Y|^{-\frac{1}{2}} \sum_{a \in Y} f_{im}(a)$. When each generator $\varepsilon' s'_k$ of $Y_k^\perp$ satisfies $m | s'_k$, that is, each generator $\varepsilon' s_k$ of Y_k satisfies $m | \frac{H'^2}{s_k}$, it reduces to the following :

$$H'^{(*H)^2}\left(\sqrt{\frac{m}{2}}\frac{1+i^{\frac{2H'^2}{m}}}{1+i}\right)^{(*H)^2} \sum_{b \in Y^\perp}\exp\left(i\pi\frac{1}{m}\sum_{k \in L} b(k)^2\right) = \prod_{k \in L} s_k \sum_{a \in Y}\exp\left(-im\pi \sum_{k \in L} a(k)^2\right)(78)$$

for a positive m, and

$$H'^{(^*H)^2} \left(\sqrt{\frac{-m}{2} \frac{1 + (-i)^{\frac{2H'^2}{-m}}}{1 - i}} \right)^{(^*H)^2} \sum_{b \in Y^\perp} \exp(i\pi \frac{1}{m} \sum_{k \in L} b(k)^2)$$

$$= \prod_{k \in L} s_k \sum_{a \in Y} \exp(-im\pi \sum_{k \in L} a(k)^2) \qquad (79)$$

for a negative m.

If $s_k = H'$ and $m | H'$, then

$$\left(\sqrt{\frac{m}{2} \frac{1 + i^{\frac{2H'^2}{m}}}{1 + i}} \right)^{(^*H)^2} \sum_{b \in Y^\perp} \exp(i\pi \frac{1}{m} \sum_{k \in L} b(k)^2) = \sum_{a \in Y} \exp(-im\pi \sum_{k \in L} a(k)^2) \qquad (80)$$

for a positive m, and

$$\left(\sqrt{\frac{-m}{2} \frac{1 + (-i)^{\frac{2H'^2}{-m}}}{1 - i}} \right)^{(^*H)^2} \sum_{b \in Y^\perp} \exp(i\pi \frac{1}{m} \sum_{k \in L} b(k)^2) = \sum_{a \in Y} \exp(-im\pi \sum_{k \in L} a(k)^2) \qquad (81)$$

for a negative m, that is,

$$\left(\sqrt{m} \exp(-i\frac{\pi}{4}) \right)^{(^*H)^2} \sum_{b \in Y^\perp} \exp(i\pi \frac{1}{m} \sum_{k \in L} b(k)^2) = \sum_{a \in Y} \exp(-im\pi \sum_{k \in L} a(k)^2) \qquad (82)$$

for a positive m, and

$$\left(\sqrt{-m} \exp(i\frac{\pi}{4}) \right)^{(^*H)^2} \sum_{b \in Y^\perp} \exp(i\pi \frac{1}{m} \sum_{k \in L} b(k)^2) = \sum_{a \in Y} \exp(-im\pi \sum_{k \in L} a(k)^2) \qquad (83)$$

for a negative m.

3.3. Poisson summation formula for Definition 2.2.3

Poisson summation formula of finite group is extended to the double infinitesimal Fourier transformation for Definition 2.2.3 on the space of functionals originally defined in [8].

Formulation

The following theorem for Definition 2.2.3 is obtained as the argument in the section 3.2.

Theorem 3.3.1. Let Y be an internal subgroup of X. Then the following is obtained, for $f \in A$,

$$|Y^{\perp \varepsilon}|^{-\frac{1}{2}} \sum_{b \in Y^{\perp \varepsilon}} (Ff)(b) = |Y|^{-\frac{1}{2}} \sum_{a \in Y} f(a) \tag{84}$$

where $< a, b >_{\varepsilon} := {}^{\star}\varepsilon \sum_{k \in L} a(k) b(k)$ and $Y^{\perp \varepsilon} := \{b \in X \mid \exp(2\pi i < a, b >_{\varepsilon}) = 1 \text{ for } \forall a \in Y\}$.

Lemma 3.3.2. $|Y^{\perp \varepsilon}| = \frac{|X|}{|Y|}$.

Proof of Lemma 3.3.2. For $k \in L$, it is denoted $Y_k := \{a(k) \in L' \mid a \in Y\}$.

$b \in Y^{\perp \varepsilon} \iff \forall a \in Y, \exp(2\pi i {}^{\star}\varepsilon \sum_{k \in L} a(k) b(k)) = 1$

$\iff \forall k \in L, {}^{\star}\varepsilon b(k) \in Y_k^{\perp}$.

For $k \in L$, generators defined by the following are written as m, n :

$Y_k = < \varepsilon' m >, \{b(k) \in L' \mid {}^{\star}\varepsilon b(k) \in Y_k^{\perp}\} = < \varepsilon' n >$.

Now

$$\exp(2\pi i {}^{\star}\varepsilon \varepsilon' m \varepsilon' n) = 1 \iff {}^{\star}\varepsilon \varepsilon' m \varepsilon' n = 1. \tag{85}$$

It is written $Y_k^{\perp \varepsilon} := \{b(k) \in L' \mid {}^{\star}\varepsilon b(k) \in Y_k^{\perp}\}$. Then $|Y_k^{\perp \varepsilon}| = m$. This is equal to $\frac{{}^{\star}HH'^2}{{}^{\star}HH'^2/m} = \frac{|L'|}{|Y_k|}$. Hence

$$|Y^{\perp \varepsilon}| = \prod_{k \in L} |Y_k^{\perp \varepsilon}| = \frac{|X|}{|Y|}. \tag{86}$$

Proof of Theorem 3.3.1.

$$|Y^{\perp \varepsilon}|^{-\frac{1}{2}} \sum_{b \in Y^{\perp \varepsilon}} (Ff)(b) = |Y^{\perp \varepsilon}|^{-\frac{1}{2}} \sum_{a \in X} \varepsilon_0 \left(\sum_{b \in Y^{\perp \varepsilon}} \exp(-2\pi i < a, b >_{\varepsilon}) \right) f(a). \tag{87}$$

Since $\sum_{b \in Y^{\perp \varepsilon}} \exp(-2\pi i < a, b >_{\varepsilon}) = \begin{cases} 0 & (a \notin Y) \\ |Y^{\perp \varepsilon}| & (a \in Y) \end{cases}$, the above is equal to

$$|Y^{\perp \varepsilon}|^{-\frac{1}{2}} \varepsilon_0 |Y^{\perp \varepsilon}| \sum_{a \in Y} f(a) = |Y|^{-\frac{1}{2}} \sum_{a \in Y} f(a). \tag{88}$$

The following is obtained:

Corollary 3.3.3.

(i) If each generator of Y_k is equal to 1, f is written as $\prod_{k\in L} f_k$, $f_k = {}^*(\mathrm{st}(f_k))|_{L'}$, and $\sum_{-\infty<n<\infty}\mathrm{st}(f_k)(n)$ converges, then

$$H^{\frac{H^2}{2}} st(\sum_{b\in Y^{\perp}} (Ff)(b)) = \prod_{k\in L}(\sum_{-\infty<n<\infty} st(f_k)(n)). \tag{89}$$

(ii) If each generator of Y_k is infinitesimal, f is written as $\prod_{k\in L} f_k$, $f_k = {}^*(\mathrm{st}(f_k))|_{L'}$, and $\mathrm{st}(f_k)$ is L_1-integrable on $\mathbf{R}$, then

$$H^{\frac{H^2}{2}} st(\sum_{b\in Y^{\perp}} (Ff)(b)) = \prod_{k\in L} \int_{-\infty}^{\infty} st(f_k)(t)dt. \tag{90}$$

Examples Theorem 3.3.1 is applied to the following two functionals :

$$1.g_i(a) = \exp(-i\pi{}^\star\varepsilon \sum_{k\in L} a(k)^2) \tag{91}$$

$$2.g_\xi(a) = \exp(-\xi\pi{}^\star\varepsilon \sum_{k\in L} a(k)^2) \tag{92}$$

where $\xi \in \mathbf{C}$, $\mathrm{Re}(\xi) > 0$. The infinitesimal Fourier transforms are :

$$1.(Fg_i)(b) = (-1)^{\frac{H}{2}} \overline{g_i(b)} \cdots (\sharp_4) \tag{93}$$

$$2.(Fg_\xi)(b) = C_\xi(b) g_\xi(\frac{b}{\xi}) \tag{94}$$

hence the following formulas are obtained :

$$1.|Y^{\perp\varepsilon}|^{-\frac{1}{2}}(-1)^{\frac{H}{2}} \sum_{b\in Y^{\perp\varepsilon}} \overline{g_i(b)} = |Y|^{-\frac{1}{2}} \sum_{a\in Y} g_i(a) \tag{95}$$

$$2.|Y^{\perp\varepsilon}|^{-\frac{1}{2}} \sum_{b\in Y^{\perp\varepsilon}} C_\xi(b) g_\xi(\frac{b}{\xi}) = |Y|^{-\frac{1}{2}} \sum_{a\in Y} g_\xi(a). \tag{96}$$

These are written as the following, explicitly :

$$1.|Y^{\perp\varepsilon}|^{-\frac{1}{2}}(-1)^{\frac{H}{2}} \sum_{b\in Y^{\perp\varepsilon}} \exp(-i\pi{}^\star\varepsilon \sum_{k\in L} b(k)^2) = |Y|^{-\frac{1}{2}} \sum_{a\in Y} \exp(-i\pi{}^\star\varepsilon \sum_{k\in L} a(k)^2) \tag{97}$$

$$2.|Y^{\perp\varepsilon}|^{-\frac{1}{2}} \sum_{b\in Y^{\perp\varepsilon}} C_\xi(b) \exp(-\frac{1}{\xi}\pi{}^\star\varepsilon \sum_{k\in L} a(k)^2) = |Y|^{-\frac{1}{2}} \sum_{a\in Y} \exp(-\xi\pi{}^\star\varepsilon \sum_{k\in L} a(k)^2). \tag{98}$$

Corollaly 3.3.3 implies the following proposition 3.3.4.

Proposition 3.3.4.

(i) If each generator of Y_k is equal to 1, then the standard parts are :

$$1.H^{\frac{H^2}{2}}(-1)^{\frac{H}{2}}st(\sum_{b\in Y_\varepsilon^\perp}\exp(-i\pi\varepsilon\sum_{k\in L}b(k)^2)) = (\sum_{-\infty<n<\infty}\exp(-i\pi\varepsilon n^2))^{H^2} \tag{99}$$

$$2.H^{\frac{H^2}{2}}st(\sum_{b\in Y_\varepsilon^\perp}C_\zeta(b)\exp(-\frac{1}{\zeta}\pi\varepsilon\sum_{k\in L}b(k)^2)) = (\sum_{-\infty<n<\infty}\exp(-\zeta\pi\varepsilon n^2))^{H^2} \tag{100}$$

$$\left(= (\theta(i\zeta))^{H^2}\right).$$

(ii) If each generator of Y_k is equal to a natural number m_k, then

$$1.H^{\frac{H^2}{2}}(-1)^{\frac{H}{2}}st(\sum_{b\in Y_\varepsilon^\perp}\exp(-i\pi\varepsilon\sum_{k\in L}b(k)^2)) = \prod_{k\in L}(m_k\sum_{-\infty<n<\infty}\exp(-i\pi\varepsilon m_k^2 n^2)) \tag{101}$$

$$2.H^{\frac{H^2}{2}}st(\sum_{b\in Y_\varepsilon^\perp}C_\zeta(b)\exp(-\frac{1}{\zeta}\pi\varepsilon\sum_{k\in L}b(k)^2)) = \prod_{k\in L}(m_k\sum_{-\infty<n<\infty}\exp(-\zeta\pi\varepsilon m_k^2 n^2)) \tag{102}$$

$$\left(= \prod_{k\in L}(m_k\theta(im_k^2\zeta))\right).$$

(iii) If each generator of Y_k is infinitesimal, then

$$2.st(\sum_{b\in Y_\varepsilon^\perp}C_\zeta(b)\exp(-\frac{1}{\zeta}\pi\varepsilon\sum_{k\in L}b(k)^2)) = (\int_{-\infty}^{\infty}\exp(-\zeta\pi t^2)dt)^{H^2} \tag{103}$$

$$\left(= \left(*\left(\frac{1}{\sqrt{\zeta}}\right)\right)^{H^2}\right).$$

The above formulation ($\sharp_4$) of $g_i(a)$ is extended to $g_{im}(a) = \exp(-im\pi{}^*\varepsilon\sum_{k\in L}a^2(k))$, for an integer m so that $m|2{}^*HH'^2$. If $m|\frac{b(k)}{\varepsilon'}$ for an arbitrary $k \in L$, it is recalled

$(Fg_{im})(b) = C_{im}(b)g_{\frac{1}{im}}(b)$, where $C_{im}(b) = \left(\sqrt{\frac{m}{2}}\frac{1+i\frac{2{}^*HH'^2}{m}}{1+i}\right)^{{}^*H^2}$ for a positive m and

$C_{im}(b) = \left(\sqrt{\frac{-m}{2}}\frac{1+(-i)\frac{2{}^*HH'^2}{-m}}{1-i}\right)^{{}^*H^2}$ for a negative m.

Hence $|Y^{\perp\varepsilon}|^{-\frac{1}{2}} \sum_{b\in Y^{\perp}} C_{im}(b) g_{\frac{1}{im}}(b) = |Y|^{-\frac{1}{2}} \sum_{a\in Y} g_{im}(a)$. When each generator $\varepsilon's'_k$ of $Y_k^{\perp\varepsilon}$ satisfies $m|s'_k$, that is, each generator $\varepsilon's_k$ of Y_k satisfies $m|\frac{{}^*HH'^2}{s_k}$, it reduces to the following:

$$H^{\frac{H^2}{2}} H'^{(*H)^2} \left(\sqrt{\frac{m}{2}} \frac{1+i^{\frac{2*HH'^2}{m}}}{1+i}\right)^{(*H)^2} \sum_{b\in Y^{\perp\varepsilon}} \exp(i\pi\frac{1}{m}{}^*\varepsilon \sum_{k\in L} b(k)^2)$$

$$= \prod_{k\in L} s_k \sum_{a\in Y} \exp(-im\pi\,{}^*\varepsilon \sum_{k\in L} a(k)^2) \qquad (104)$$

for a positive m, and

$$H^{\frac{H^2}{2}} H'^{(*H)^2} \left(\sqrt{\frac{-m}{2}} \frac{1+(-i)^{\frac{2*HH'^2}{-m}}}{1-i}\right)^{(*H)^2} \sum_{b\in Y^{\perp\varepsilon}} \exp(i\pi\frac{1}{m}{}^*\varepsilon \sum_{k\in L} b(k)^2)$$

$$= \prod_{k\in L} s_k \sum_{a\in Y} \exp(-im\pi\,{}^*\varepsilon \sum_{k\in L} a(k)^2) \qquad (105)$$

for a negative m. If $s_k = H'$ and $m|H'$, then

$$H^{\frac{H^2}{2}} \left(\sqrt{\frac{m}{2}} \frac{1+i^{\frac{2*HH'^2}{m}}}{1+i}\right)^{(*H)^2} \sum_{b\in Y^{\perp\varepsilon}} \exp(i\pi\frac{1}{m} \sum_{k\in L} b(k)^2) = \sum_{a\in Y} \exp(-im\pi\,{}^*\varepsilon \sum_{k\in L} a(k)^2) \quad (106)$$

for a positive m, and

$$H^{\frac{H^2}{2}} \left(\sqrt{\frac{-m}{2}} \frac{1+(-i)^{\frac{2*HH'^2}{-m}}}{1-i}\right)^{(*H)^2} \sum_{b\in Y^{\perp\varepsilon}} \exp(i\pi\frac{1}{m} \sum_{k\in L} b(k)^2)$$

$$= \sum_{a\in Y} \exp(-im\pi\,{}^*\varepsilon \sum_{k\in L} a(k)^2) \qquad (107)$$

for a negative m, that is,

$$H^{\frac{H^2}{2}} \left(\sqrt{m}\exp(-i\frac{\pi}{4})\right)^{(*H)^2} \sum_{b\in Y^{\perp\varepsilon}} \exp(i\pi\frac{1}{m} \sum_{k\in L} b(k)^2) = \sum_{a\in Y} \exp(-im\pi\,{}^*\varepsilon \sum_{k\in L} a(k)^2) \quad (108)$$

for a positive m, and

$$H^{\frac{H^2}{2}}\left(\sqrt{-m}\exp(i\frac{\pi}{4})\right)^{(^*H)^2}\sum_{b\in Y^{\perp_\varepsilon}}\exp(i\pi\frac{1}{m}\sum_{k\in L}b(k)^2)=\sum_{a\in Y}\exp(-im\pi{}^*\varepsilon\sum_{k\in L}a(k)^2) \quad (109)$$

for a negative m.

4. Quantum field theory and Zeta function

In this section the quantum field theory is developed by using the double infinitesimal Fourier transform. The propagator for a system of the harmonic oscillators is considered in the quantum field theory.

4.1. Path integral in the quantum field theory

Definition4.1.1. A path integral of $f(\in A)$ is defined as follows:

$$\sum_{a\in X}\varepsilon_0 f(a) \quad (110)$$

with $\varepsilon_0:=(H')^{-(^*H)^2}\in{}^*(^*\mathbf{R})$.

It is briefly explained that the complexification of the propagator for the harmonic oscillator is represented as the following path integral. In Feynman's formulation of quantum mechanics([2]), the propagator of the one-dimensional harmonic oscillator is the following path integral: $K(q,q_0,t)$

$$=\lim_{n\to\infty}\int_{\mathbf{R}^n}(\frac{m}{2\pi i\hbar})^{(n+1)/2}\exp\left(\frac{i\epsilon}{\hbar}\sum_{j=1}^{n+1}(\frac{m}{2}(\frac{x_j-x_{j-1}}{\epsilon})^2-\frac{m}{2}\lambda^2 x_j^2)\right)dx_1 dx_2\cdots dx_n \quad (111)$$

where $x_0=q_0$, $x_{n+1}=q$, $\epsilon=\frac{t}{n}$. In nonstandard analysis, it is known that, for a sequence a_n,

$$\lim_{n\to\infty}a_n=a\quad iff\quad {}^*a_w\approx a \quad (112)$$

for any infinite natural number $\omega\in{}^*\mathbf{N}-\mathbf{N}$, where $^*a_n$ is the $*$ extension of $\{a_n\}_{n\in N}$, and $\approx$ means that $^*a_w-a$ is infinitesimal, that is, the standard part of a_w is a, usually denoted by $st(^*a_w)=a$. The standard part of the nonstandard path integral is written as

$$st\int_{^*\mathbf{R}^w}(\frac{m}{2\pi i\hbar})^{(\omega+1)/2}\exp\left(\frac{i\epsilon}{\hbar}\sum_{j=1}^{w+1}(\frac{m}{2}(\frac{x_j-x_{j-1}}{\epsilon})^2-\frac{m}{2}\lambda^2 x_j^2)\right)dx_1 dx_2\cdots dx_w. \quad (113)$$

By extending t to a complex number, the path integral is complexified to the following :

$$st \int_{*\mathbf{R}^w} (\frac{m}{2\pi i \hbar \epsilon})^{(w+1)/2} \exp\Big(\frac{i\epsilon}{\hbar} \sum_{j=1}^{w+1} (\frac{m}{2}(\frac{x_j - x_{j-1}}{\epsilon})^2 - \frac{m}{2}\lambda^2 x_j^2)\Big) dx_1 dx_2 \cdots dx_w. \qquad (114)$$

Theorem 4.1.2. Let $t \in \mathbf{R}$, $t \neq st(\pm\frac{\sqrt{2}n}{\lambda}\sqrt{1 - \cos(\frac{k\pi}{\omega+1})})$, $k = 1, 2, \cdots, \omega$, or for $t \in \mathbf{C}$, whose imaginary part is negative. The complexified one-dimensional harmonic oscillator standard functional integral is given by $(\frac{m}{2\pi i \hbar})^{\frac{1}{2}} \sqrt{\frac{\lambda}{\sin(\lambda t)}} \exp(\frac{im}{\hbar} \frac{\lambda}{\sin \lambda t}((q_0^2 + q^2)\cos \lambda t - 2qq_0))$.

Proof. If $t \neq st(\pm\frac{\sqrt{2}n}{\lambda}\sqrt{1 - \cos(\frac{k\pi}{\omega+1})})$, $k = 1, 2, \cdots, \omega$, then

$$t \neq \pm\frac{\sqrt{2}n}{\lambda}\sqrt{1 - \cos(\frac{k\pi}{\omega+1})}, k = 1, 2, \cdots, \omega \qquad (115)$$

for arbitrary infinite number ω. The theorem is followed from the discrete calculation using the matrix representation of the operator(cf. [7]).

It corresponds to the well-known real propagator for one dimensional harmonic oscillator. For the d-dimensional harmonic oscillator, d-dimensional vectors are written as $\mathbf{q_0}, \mathbf{q}$, the square norms are $|\mathbf{q_0}|^2, |\mathbf{q}|^2$, and the inner product of $\mathbf{q_0}, \mathbf{q}$ is $\mathbf{q_0 q}$. We have :

Corollary 4.1.3. For the complexified d-dimensional harmonic oscillator standard functional integral, the complexified propagator is given by

$$(\frac{m}{2\pi i \hbar})^{\frac{d}{2}}(\frac{\lambda}{\sin(\lambda t)})^{\frac{d}{2}} \exp(\frac{im}{\hbar} \frac{\lambda}{\sin \lambda t}((|\mathbf{q_0}|^2 + |\mathbf{q}|^2)\cos \lambda t - 2\mathbf{q q_0})). \qquad (116)$$

Proof. By factorizing Theorem 4.1.2 into a product on d dimensional, the corollary is obtained.

The trace of the compiexified propagator is calculated for one dimensional harmonic oscillator.

Since

$$\int_{-\infty}^{\infty} (\frac{m}{2\pi i \hbar})^{\frac{1}{2}} \sqrt{\frac{\lambda}{\sin(\lambda t)}} \exp(\frac{im}{\hbar} \frac{\lambda}{\sin \lambda t} 2((\cos \lambda t - 1)q^2)) dq = \frac{1}{2i \sin(\lambda t/2)} \qquad (117)$$

the following is obtained (cf.[6],[7]):

Theorem 4.1.4. Let $t \in \mathbf{R}$, $t \neq \pm st(\frac{\sqrt{2}\omega}{\lambda}\sqrt{1 - \cos(\frac{k\pi}{\omega+1})})$, $k = 1, 2, \cdots, \omega$, or $t \in \mathbf{C}$, whose imaginary part is negative. The trace of the complexified one-dimensional harmonic oscillator standard functional integral is given by $\frac{1}{2i \sin(\lambda t/2)}$.

Proof. By putting $q_0 = q$ in $(\frac{m}{2\pi i\hbar})^{\frac{1}{2}}\sqrt{\frac{\lambda}{\sin(\lambda t)}}\exp(\frac{im}{\hbar}\frac{\lambda}{\sin\lambda t}((q_0^2+q^2)\cos\lambda t - 2qq_0))$, the trace is the following integral :

$$\int_{-\infty}^{\infty}(\frac{m}{2\pi i\hbar})^{\frac{1}{2}}\sqrt{\frac{\lambda}{\sin(\lambda t)}}\exp(\frac{im}{\hbar}\frac{\lambda}{\sin\lambda t}2((\cos\lambda t - 1)q^2))dq = \frac{1}{2i\sin(\lambda t/2)}. \qquad (118)$$

Corollary 4.1.5. If the potential is modified to $V(q) = \frac{m}{2}(\frac{\lambda^2}{2}|q|^2 - \frac{\lambda}{2})$, then the trace is $\frac{1}{2i\sin(\lambda t/2)}\exp(\frac{\lambda t}{2}))$.

For the d-dimensional harmonic oscillator, the following is obtained :

Corollary 4.1.6. For the trace of the modified complexified propagator for d-dimensional harmonic oscillator, the trace is $(\frac{1}{2i\sin(\lambda t/2)}\exp(\frac{\lambda t}{2})))^d$.

In the next section, Corollaries 4.1.5 and 4.1.6 are used to treat an infinite dimensional harmonic oscillator.

4.2. Representation of the zeta function.

Corollary 4.1.5 is extended to an infinite dimensional harmonic oscillator using nonstandard analysis. For it the three types of extension $^*\mathbf{R}$,$^{**}\mathbf{R}$, $^{\#**}\mathbf{R}$ of $\mathbf{R}$ are prepared corresponding to Definition 2.2.2 , then the three stages of infinite numbers exist. In these three extension fields, we fix infinite natural numbers $H_F \in {}^*\mathbf{N}$,$H_T \in {}^{**}\mathbf{N}$, $H'' \in 2^{\#**}\mathbf{N}$. Let T be a positive standard real number and let ϵ_T , ϵ'' be infinitesimals in $^{**}\mathbf{R}$, $^{\#**}\mathbf{R}$ defined by $\frac{T}{H_T}$, $\frac{1}{H''}$. A lattice L'' and two function space X, A are defined as the following:

$$L'' := \left\{\epsilon''z'' \,\middle|\, z'' \in {}^{\#**}\mathbf{Z}, -\frac{H''}{2} \leq {}'''z'' < \frac{H''}{2}\right\}).$$

$$X := \{\alpha :^{\#*} \{0,1,\cdots.H_F - 1\} \to L'', internal\},$$

$$A := \{a :^{\#} \{0,1,\cdots.H_T\} \to X, internal\}.$$

Then an element a of A is written as the component $(a_j^k, 0 \leq j \leq H_T, 0 \leq k \leq H_F - 1)$. All prime numbers are ordered as $p(1) = 2, p(2) = 3, \dots$, $p(n) < p(n+1), \dots$, that is, p is a mapping from $\mathbf{N}$ to the set of prime numbers, $p : \mathbf{N} \to \{prime\ number\}$. Let λ_k be $\ln^* p(k)$ for each k , $0 \leq k \leq H_F - 1$. A potential $V_k :^{\#**}\mathbf{R} \to^{\#**}\mathbf{R}$ is defined for each k , $0 \leq k \leq H_F - 1$, as $V_k(q) = \frac{\lambda_k^2}{2}|q|^2 - \frac{\lambda_k}{2}$. An element α of X is written as the component $\alpha = (\alpha^k, 0 \leq k \leq H_F - 1)$.

Let V be a global potential as the following:

$$V(\alpha) = \sum_{k=0}^{H_F-1} V_k(\alpha^k) \left(= \sum_{k=0}^{H_F-1}\left(\frac{\lambda_k^2}{2}|\alpha^k|^2 - \frac{\lambda_k}{2}\right)\right). \qquad (119)$$

In order to transport t to later, the element $-\frac{\lambda_k}{2}$ is put in the usual potential for harmonic oscillators. It is considered the following summation $K(a,b,t)$ depending of $a,b \in X$:

$$K(a,b,t) = \sum_{a \in A, a_0=a.a_{H_T}=b} ((\epsilon')^{H_F})^{H_T}((\frac{1}{2\pi\epsilon_T})^{H_F})^{H_T} \exp(\epsilon_T(\frac{1}{2}\sum_{j=1}^{H_T}|\frac{a_j - a_{j-1}}{\epsilon_T}|^2 - V(a_j))).(120)$$

Then $K(a,b,t)$ is calculated,

$$K(a,b,t) = (\frac{1}{2\pi\epsilon_T})^{H_T} \prod_{k=1}^{H_F-1} \sum_{a_j^k \in L', 0 \le j \le H_T-1} (\epsilon')^{H_T}(\frac{1}{2\pi\epsilon_T})^{H_T} \exp(\epsilon_T(\frac{1}{2}\sum_{j=1}^{H_T}|\frac{a_j^k - a_{j-1}^k}{\epsilon_T}|^2 - V(a_j^k))), \quad (121)$$

where $a_0 = a, a_{H_T} = b$.

The summation $\sum_{a \in X}(\epsilon')^{H_T}K(a,a.t)$ is denoted by $tr(K(a,a,t))$. Three correspondences putting standard parts are written as $st_\# : {}^{\#**}\mathbf{R} \to {}^{**}\mathbf{R}$, $st_\star : {}^{**}\mathbf{R} \to {}^{*}\mathbf{R}$, $st_* : {}^{*}\mathbf{R} \to \mathbf{R}$. When there are no confusion, they are simply written as st. The composition $st_* \circ st_\star \circ st_\# : {}^{\#**}\mathbf{R} \to \mathbf{R}$ is denoted also by st for simplicity.

Theorem 4.2.1. If the real part of t is greater than 1, the standard part $st(tr(K(a,a,t)))$ of $tr(K(a,a,t))$ corresponds to Riemann's zeta function $\zeta(t)$.

Proot. The standard parts of $tr(K(a,a,t))$ as follows.

$$st_\#(tr(K(a,a,t))) =$$

$$\prod_{k=1}^{H_F-1} \int\int \cdots \int (\frac{1}{2\pi\epsilon_T})^{H_T} \exp(\epsilon_T(\frac{1}{2}\sum_{j=1}^{H_T}(\frac{q_j^k - q_{j-1}^k}{\epsilon_T})^2 - V(q_j^k)))dq_0^k dq_1^k \cdots dq_{H_T-1}^k \quad (122)$$

$$= \prod_{k=1}^{H_F-1} \int\{\int \cdots \int (\frac{1}{2\pi\epsilon_T})^{H_T} \exp(\epsilon_T(\frac{1}{2}\sum_{j=1}^{H_T}(\frac{q_j^k - q_{j-1}^k}{\epsilon_T})^2 - V(q_j^k)))dq_1^k \cdots dq_{H_T-1}^k\}dq_0^k \quad (123)$$

by Fubini's theorem. Furthermore,

$$st_\star st_\#(tr(K(a,a,t))) =$$

$$= \prod_{k=1}^{H_F-1} st_\star\{\int\{\int \cdots \int (\frac{1}{2\pi\epsilon_T})^{H_T} \exp(\epsilon_T(\frac{1}{2}\sum_{j=1}^{H_T}(\frac{q_j^k - q_{j-1}^k}{\epsilon_T})^2 - V(q_j^k)))dq_1^k \cdots dq_{H_T-1}^k\}dq_0^k\} \quad (124)$$

by the same calculation of Theorem 4.1.2 (cf.[6],[7]) ,

$$= \prod_{k=0}^{H_F-1} \left(\frac{1}{2i \sin(\frac{\lambda_k t}{2i})} \exp(\frac{\lambda_k t}{2}) \right) = \prod_{k=0}^{H_F-1} \frac{1}{1 - p_k^{-t}}. \tag{125}$$

By Lebesque's convergence theorem, $(st_*(st_*(st_\#(tr(K(a,a,t)))))) = \prod_{k=0}^{\infty} \frac{1}{1-p_k^{-t}} = \zeta(t)$, if the real part of t is positive.

4.2. Another representation of the zeta function

In this section, both st_*, st_* and $st_\#$ are denoted as st for the simplification. A functional is defined on X, and a relationship between the functional and Riemann's zeta function is shown later. The nonstandard extension $*p : *\mathbf{N} \to *\{\text{prime number}\}$ is written as $*p([l_\mu]) = [p(l_\mu)]$, and a mapping $\tilde{p} : *\mathbf{N} \to *(*\{\text{prime number}\})$ is defined as $\tilde{p}([l_\mu]) = *[p(l_\mu)]$. For $s \in \mathbf{C}$, $Z_s(\in A)$ is defined as the following :

$$Z_s(a) := \prod_{k \in L} \tilde{p}\left(H\left(k + \frac{H}{2}\right) + 1\right)^{(-s(a(k) + \frac{H'}{2}))}. \tag{126}$$

Now $H(k + \frac{H}{2}) + 1$ is an element of $*\mathbf{N}$ and $a(k) + H'/2$ is an element of $*(*\mathbf{N})$. Then $Z_s(a)$ is calculated as $\exp(-s \sum_{k \in L} \log(\tilde{p}(H(k + \frac{H}{2}) + 1))a(k)) \prod_{k \in L} \tilde{p}(H(k + \frac{H}{2}) + 1)^{-s\frac{H'}{2}}$. The following theorem is obtained for the Fourier transform of Z_s for Definition 2.2 1:

Theorem 4.3.1.

$$(F((Z_s))(b) = \left(\prod_{k \in L} \tilde{p}\left(H\left(k + \frac{H}{2}\right) + 1\right) \right)^{-s\frac{H'}{2}}$$

$$\cdot \prod_{k \in L} \varepsilon' \frac{\sinh((2\pi i b(k) + s \log \tilde{p}(H(k + \frac{H}{2}) + 1))\frac{H'}{2})}{\exp(-\frac{\varepsilon'}{2}(2\pi i b(k) + s \log \tilde{p}(H(k + \frac{H}{2}) + 1))\sinh(\frac{\varepsilon'}{2}(2\pi i b(k) + s \log \tilde{p}(H(k + \frac{H}{2}) + 1))}. \tag{127}$$

Proof.

$$(F((Z_s))(b) = \left(\prod_{k \in L} \tilde{p}\left(H\left(k + \frac{H}{2}\right) + 1\right) \right)^{-s\frac{H'}{2}}$$

$$\cdot \sum_{a \in X} \varepsilon_0 \exp(-s \sum_{k \in L} \log \tilde{p}(H(k + \frac{H}{2}) + 1)a(k)) \exp(-2\pi i \sum_{k \in L} a(k)b(k))$$

$$= \left(\prod_{k \in L} \tilde{p}\left(H\left(k + \frac{H}{2}\right) + 1\right) \right)^{-s\frac{H'}{2}} \cdot \sum_{a \in X} \varepsilon_0 \exp(-(2\pi i\, b(k) + s \log \tilde{p}(H(k + \frac{H}{2}) + 1))a(k))$$

$$= \left(\prod_{k \in L} \tilde{p}\left(H\left(k + \frac{H}{2}\right) + 1\right) \right)^{-s\frac{H'}{2}}$$

$$\cdot \prod_{k\in L}\varepsilon' \frac{\sinh((2\pi i\, b(k)+s\log \tilde{p}(H(k+\frac{H}{2})+1))\frac{H'}{2})}{\exp(-\frac{\varepsilon'}{2}(2\pi i\, b(k)+s\log \tilde{p}(H(k+\frac{H}{2})+1))\sinh(\frac{\varepsilon'}{2}(2\pi i\, b(k)+s\log \tilde{p}(H(k+\frac{H}{2})+1))}.$$

Riemann's zeta function $\zeta(s)$ is defined by $\zeta(s)=\prod_{l=1}^{\infty}\frac{1}{1-p(l)^{-s}}$ for $\mathrm{Re}(s)>1$. Let Y_Z be a subgroup of X so that each generator of $(Y_Z)_k$ is equal to 1. Then the following theorem is obtained :

Theorem 4.3.2. If $\mathrm{Re}(s)>1$, then $\mathrm{st}(\mathrm{st}(\sum_{a\in Y_Z}(Z_s))(a)))=\zeta(s)$.

Proof. $\mathrm{st}(\mathrm{st}(\sum_{a\in Y_Z}(Z_s)(a)))=\mathrm{st}\left(\mathrm{st}\left(\left(\prod_{k\in L}\tilde{p}(H(k+\frac{H}{2})+1)\right)^{(-s(a(k)+\frac{H'}{2}))}\right)\right)$

$$=\mathrm{st}\left(\mathrm{st}\left(\prod_{k\in L}\frac{1-\tilde{p}(H(k+\frac{H}{2})+1)^{-sH'}}{1-\tilde{p}(H(k+\frac{H}{2})+1)^{-s}}\right)\right)=\mathrm{st}\left(\prod_{k\in L}\frac{1}{1-\tilde{p}(H(k+\frac{H}{2})+1)^{-s}}\right)=\zeta(s). \quad (128)$$

Furthermore, Poisson summation formula and Theorem 4.3.2 imply the following :

Corollary 4.3.3.

$$\mathrm{st}(\sum_{b\in Y_{\bar{Z}}^{\perp}}(F(Z_s)(b))=\mathrm{st}\left(\prod_{k\in L}\frac{1-\tilde{p}(H(k+\frac{H}{2})+1)^{-sH'}}{1-\tilde{p}(H(k+\frac{H}{2})+1)^{-s}}\right). \quad (129)$$

Hence we obtain :

$$\mathrm{st}(\mathrm{st}(\sum_{b\in Y_{\bar{Z}}^{\perp}}(F(Z_s)(b))))=\zeta(s) \quad (130)$$

for $\mathrm{Re}(s)>1$.

In general, the physical theory has variables for position, time, and fields. Especially there are many kinds of variables in quantum field theory. The function depends on such variables mixed as $f(\mathbf{q},t,\mathbf{a},\mathbf{b},\mathbf{c})$ where $\mathbf{q}$ is position, t is time, $\mathbf{a},\mathbf{b},\mathbf{c}$ are fields. When the function is treated for such mixed variables, the Kinoshita's infinitesimal Fourier transform and our double infinitesimal Fourier transform are applied in the double extended number field. The two kinds of Fourier transforms can be used for one function. In the theory, the delta functions for variable and for fields have different infinitesimals and infinite values. The delta function for fields has an infinitesimal much smaller and much bigger infinite number. However they can be treat in the double extended number field. Two kinds of delta functions are defined with another degrees. One delta function has an infinitesimal of the first degree and the other delta function has an infinitesimal of the second degree. The infinitesimal of the second one can not be observable with respect to the first one.

5. Conclusion

The real and complex number fields are extended to the larger number fields where there are many infinitesimal and infinite numbers. A lattice of infinitesimal width is included in the extended real number field. An infinitesimal Fourier transform theory is constructed on the infinitesimal lattice. These extended number fields are furthermore extended to much higher generalized fields where there exist much higher infinitesimal and infinite numbers. A double infinitesimal Fourier transform theory is developed on these double extended number fields. The usual formulae for Fourier theory are satisfied in the theory, especially the Poisson summation formula. The Fourier theory is based on the integral theory for functionals corresponding to the path integral in the physics. The theory is associated to the physical theory in the quantum field theory which is mathematically rigorous . For an application for the double infinitesimal calculation, Riemann's zeta function is represented as such an integral for the propagator of an infinite dimensional harmonic oscillator.

Acknowledgement

Supported by JSPS Grant-in-Aid for Scientific Research (B) No. 25287010.

Author details

Takashi Gyoshin Nitta

Department of Mathematics, Faculty of Education, Faculty of Education, Mie University, Japan

References

[1] Dirac PAM. The principles of quantum mechanics. Oxford Univ.Press; 1930.

[2] Feynman RP, Hibbs AR. Quantum mechanics and path integrals. McGrow-Hill Inc. All rights; 1965.

[3] Gordon EI. Nonstandard methods in commutative harmonic analysis, Translations of mathematical monographs 164 American mathematical society;1997.

[4] Kinoshita M. Nonstandard representation of distribution I. Osaka J. Math.1988; 25 805-824.

[5] Kinoshita M. Nonstandard representation of distribution II. Osaka J. Math. 1990; 27 843-861.

[6] Nitta T. Complexification of the propagator for the harmonic oscillator. Topics in contemporary differential geometry, complex analysis and mathematical physics; 2007. p261-268.

[7] Nitta T. A complexified path integral for a system of harmonic oscillators. Nonlinear Anal. 2009; 71 2469-2473.

[8] Nitta T, Okada T. Double infinitesimal Fourier transformation for the space of functionals and reformulation of Feynman path integral, Lecture Note Series in Mathematics, Osaka University 2002 7 255-298 in Japanese.

[9] Nitta T, Okada T. Infinitesimal Fourier transformation for the space of functionals. Nihonkai Math. J. 2005; 16 1-21.

[10] Nitta T, Okada T. Poisson summation formula for the space of f functionals, Topics in contemporary differential geometry, complex analysis and mathematical physics. ;2007. p261-268.

[11] Nitta T, Okada T, Tzouvaras A. Classification of non-well-founded sets and an application. Math. Log. Quart. ;2003 49 187-200.

[12] Nitta T,Péraire Y. Divergent Fourier analysis using degrees of observability. Nonlinear Anal. 2009; 71 2462-2468.

[13] Péraire Y. Théorie relative des ensembles internes. Osaka J.Math. 1992; 29 267-297.

[14] Péraire Y. Some extensions of the principles of idealization transfer and choice in the relative internal set theory, Arch. Math. Logic . 1995; 34 269-277.

[15] Remmert R. Theory of complex functions. Graduate Texts in Mathematics 122 Springer Berlin-Heidelberg-New York; 1992.

[16] Saito M. Ultraproduct and non-standard analysis. Tokyo tosho; 1976 in Japanese.

[17] Satake I. The temptation to algebra. Yuseisha; 1996 in Japanese.

[18] Takeuti G. Dirac space. Proc. Japan Acad. 1962; 38 414-418.

3

Gabor-Fourier Analysis

Nafya Hameed Mohammad and Massoud Amini

Additional information is available at the end of the chapter

1. Introduction

The notion of Gabor transform, named after Dennis Gabor [1], is a special case of the short-time Fourier transform. The Gabor analysis, as it stands now, is a rather new field, but the idea goes back quite some while. Dennis Gabor investigated in [1] the representation of a one dimensional signal in two dimensions, time and frequency. He suggested to represent a function by a linear combination of translated and modulated Gaussians. Interestingly, there is a tight connection between this approach and quantum mechanics (c.f. [2]). On the mathematical side, the representation of functions by other functions was further investigated, leading to the theory of atomic decompositions, developed by Feichtinger and Gröchenig [3].

Gabor transform and Gabor expansion have long been recognized as very useful tools for the signal processing, and it is because of this reason over the recent years, an increasing attention has been given to the study of them in engineering and applied Mathematics, see for instance [4, 5]. Borichev et al. [6] studied the stability problem for the Gabor expansions generated by a Gaussian function. In [7], Ascensi and Bruna proved that the unique Gabor atom with analytical Gabor space, the image of $L^2(\mathbb{R})$ under the Gabor transform, is the Gaussian function. The structure of Gabor and super Gabor spaces inside $L^2(\mathbb{R}^{2d})$ is studied by Abreu [8]. Christensen [9] has done a comprehensive study of the Gabor system and has asked for the necessary and sufficient conditions to get a frame for $L^2(\mathbb{R})$.

Today Gabor analysis and the closely related wavelet analysis are considered topics in harmonic analysis. The basic idea behind wavelet analysis is that the notion of an orthonormal basis is not always useful. Sometimes it is more important for a decomposing set to have special properties, like good time frequency localization, than to have unique coefficients. This led to the concept of frames, which was introduced by Duffin and Schaefer in [10] and was made popular by Daubechies, and today is one of the most important foundations of Gabor theory and a fundamental subject in harmonic analysis.

Most examples of Gabor frames correspond to regular nets of points. That is, sets of the type $\{e^{2\pi i bnt}h(t-am)\}_{n,m\in\mathbb{Z}^d}$. One can usually find sufficient and necessary conditions for the existence of such kind of frames, with a variety of applications. For technical reasons, however, one needs to work with frames coming from irregular grids. One of the main purposes of this chapter is to study perturbations of irregular Gabor frames and the problem of stability.

On the other hand, the theory of nonharmonic Fourier series is concerned with the completeness and expansion properties of sets of complex exponentials $\{e^{i\lambda_n t}\}$ in $L^p[-\pi,\pi]$. In 1952, Duffin and Schaeffer [10] used frames to study this theory, and later Young put together many results in his book [11]. Reid [12] proved that if $\{\lambda_n\}$ is a sequence of real numbers whose differences are nondecreasing, then the set of complex exponentials $\{e^{i\lambda_n t}\}$ is a Riesz-Fischer sequence in $L^2[-A, A]$ for every $A > 0$. Jaffard [13] investigated how the regularity of nonharmonic Fourier series is related to the spacing of their frequencies, and this is obtained by using a transformation which simultaneously captures the advantages of the Gabor and wavelet transforms.

In this chapter, we restate and prove some classical results of (nonharmonic) Fourier expansions for Gabor systems instead of sets of complex exponentials. Some of the results may be known or obtainable via Hilbert space methods, but the main advantage of this work is that it uses analytic methods and can be fully understood with elementary knowledge of functional and complex analysis in several variables [14, 15].

2. Preliminaries

Let us introduce the notions and basic results, needed later in the chapter.

Definition 2.1 We say that $\Lambda = \{z_j\}_{j\in\mathbb{N}} \subset \mathbb{C}^d$ is a *separated* set if there exists $\varepsilon > 0$ such that $|z_i - z_j| \geq \varepsilon$, $i \neq j$. The largest of such ε is called the *separation constant* of Λ. A finite union of separated sets is called a *relatively separated set*.

Definition 2.2 A sequence of vectors $\{x_n\}$ in a normed space $\mathcal{X}$ is said to be *complete* if its linear span is dense in $\mathcal{X}$, that is, if for each vector x and each $\varepsilon > 0$ there is a finite linear combination $c_1 x_1 + \cdots + c_n x_n$ such that

$$\|x - (c_1 x_1 + \cdots + c_n x_n)\| < \varepsilon.$$

Definition 2.3 A sequence $\{f_n\}$ in a Hilbert space H is said to be a *Bessel sequence* if

$$\sum_{n=1}^{\infty} |\langle f, f_n\rangle|^2 < \infty$$

for every element $f \in H$. It is called a *Riesz-Fischer sequence* if the moment problem

$$\langle f, f_n\rangle = c_n, \quad n \geq 1$$

admits at least one solution $f \in H$ whenever $\{c_n\} \in l^2$.

Proposition 2.4 Let $\{f_n\}$ be a sequence in a Hilbert space H. Then

(i) $\{f_n\}$ is a Bessel sequence with bound M if and only if the inequality

$$\left\| \sum c_n f_n \right\|^2 \leq M \sum |c_n|^2$$

holds for every finite sequence of scalars $\{c_n\}$;

(ii) $\{f_n\}$ is a Riesz-Fischer sequence with bound m if and only if the inequality

$$m \sum |c_n|^2 \leq \left\| \sum c_n f_n \right\|^2$$

holds for every finite sequence of scalars $\{c_n\}$.

Remark 2.5 For a sequence $\{f_n\}$ in a Hilbert space H, the moment problem

$$\langle f, f_n \rangle = c_n, \quad n \geq 1$$

has at most one solution for every choice of the scalars $\{c_n\}$ if and only if $\{f_n\}$ is complete.

Definition 2.6 A countable family $\{f_k\}_{k \in I}$ in a separable Hilbert space H is a *frame* for H if there exist constants A and B such that $0 < A \leq B < \infty$ and

$$A\|f\|^2 \leq \sum_{k \in I} |\langle f, f_k \rangle|^2 \leq B\|f\|^2, \quad f \in H.$$

A, B are called the *lower* and *upper frame bounds* respectively. They are not unique: the biggest lower bound and the smallest upper bound are called the *optimal frame bounds*. Every element in H has at least one representation as an infinite linear combination of the frame elements.

Definition 2.7 Let $c \in \mathbb{R}^d$, the unitary operators T_c and M_c on $L^2(\mathbb{R}^d)$ defined by $T_c f(t) = f(t - c)$ and $M_c f(t) = e^{2\pi i c t} f(t)$ are called the *Translation* and *Modulation* operator, respectively. For a discrete set $\Lambda = \{z_j\}_{j \in \mathbb{Z}}$ in $\mathbb{C}^d$ and a fixed nonzero window function $h \in L^2(\mathbb{R}^d)$, we define the *Gabor system* $G(h, \Lambda)$ as:

$$G(h, \Lambda) = \{M_y T_x h(t) = e^{2\pi i y t} h(t - x); \ x + iy \in \Lambda\}.$$

For simplicity we denote $e^{2\pi i y t} h(t - x)$ by $h_z(t)$, where $z = x + iy$. Gabor systems were first introduced by Gabor [1] in 1946 for signal processing, and is still widely used. A Gabor system is said to be *exact* in $L^2(\mathbb{R}^d)$ if it is complete, but fails to be complete on the removal of any one term.

If $G(h, \Lambda)$ is a frame for $L^2(\mathbb{R}^d)$, it is called a *Gabor frame* or *Weyl-Heisenberg frame*.

Definition 2.8 Let f be an entire function. For $r > 0$, the *maximum modulus function* is $M(r) = \max\{|f(z)| : |z| = r\}$. Unless f is a constant of modulus less than or equal to 1, its *order*, which is denoted by ρ, is defined by

$$\rho = \limsup_{r \to \infty} \frac{\log \log M(r)}{\log r}.$$

Simple examples of functions of finite order include e^z, $\sin z$, and $\cos z$, all of which are of order 1, and $\cos \sqrt{z}$, which is of order $\frac{1}{2}$. Every polynomial is of order 0; the order of a constant function is of course 0 and the function e^{e^z} is of infinite order.

Remark 2.9 An entire function has an order of growth $\leq \rho$ if $|f(z)| \leq A\, e^{B|z|^\rho}$.

The following is the fundamental factorization theorem for entire functions of finite order. It is due to Hadamard who used the result in his celebrated proof of the Prime Number Theorem. It is one of the classical theorems in function theory.

Theorem 2.10 (Hadamard Factorization Theorem) Let f be an entire function of finite order ρ, $\{z_n\}$ be the zeros of f different from 0, k be the order of zero of f at the origin, and

$$f(z) = z^k e^{g(z)} \prod_{n=1}^{\infty} (1 - \frac{z}{z_n})$$

be its canonical Factorization, then $g(z)$ is a polynomial of degree no longer than ρ.

Definition 2.11 The *(Bargmann-)Fock space*, $\mathcal{F}(\mathbb{C}^d)$, is the Hilbert space of all entire functions f on $\mathbb{C}^d$ for which

$$\|f\|_{\mathcal{F}}^2 = \int_{\mathbb{C}^d} |f(z)|^2 e^{-\pi |z|^2} dz,$$

is finite. The natural inner product on $\mathcal{F}(\mathbb{C}^d)$ is defined by

$$\langle f, g \rangle_{\mathcal{F}} = \int_{\mathbb{C}^d} f(z)\overline{g(z)} e^{-\pi |z|^2} dz; \quad f, g \in \mathcal{F}(\mathbb{C}^d).$$

The *Bargmann transform* of a function $f \in L^2(\mathbb{R}^d)$ is the function Bf on $\mathbb{C}^d$ defined by

$$Bf(z) = 2^{\frac{d}{4}} e^{-\frac{\pi}{2}z^2} \int_{\mathbb{R}^d} f(t) e^{-\pi t^2} e^{2\pi tz}\, dt.$$

Definition 2.12 Fix a function $h \in L^2(\mathbb{R}^d)$ (called the window function). The *Gabor transform* with respect to the window h is the isomorphic inclusion

$$V_h : L^2(\mathbb{R}^d) \longrightarrow L^2(\mathbb{C}^d),$$

defined by

$$V_h f(z) = \langle f(t), h_z(t) \rangle = 2^{d/4} \int_{\mathbb{R}^d} f(t)\overline{h(t-x)} e^{-2\pi i ty}\, dt$$

for every $f \in L^2(\mathbb{R}^d)$ and $z = x + iy \in \mathbb{C}^d$. The following subspace of $L^2(\mathbb{C}^d)$ which is the image of $L^2(\mathbb{R}^d)$ under the Gabor transform with the window h,

$$\mathcal{G}_h = \{V_h f : f \in L^2(\mathbb{R}^d)\},$$

is called *Gabor space* or *model space*. A simple calculation shows that the Bargmann transform is related to the Gabor transform with the Gaussian window $g(t) = 2^{d/4} e^{-\pi t^2}$ in $L^2(\mathbb{R}^d)$ by the formula

$$V_g f(x - iy) = e^{i\pi xy} e^{-\pi \frac{|x+iy|^2}{2}} (Bf)(x + iy). \tag{1}$$

For more details we refer the reader to [2, 7, 8, 17].

3. Gabor Expansion

Here we discuss the fundamental completeness properties of the Gabor systems. The most extensive results in the case of the sets of complex exponentials $\{e^{i\lambda_n t}\}$ over a finite interval of the real axis were obtained by Paley and Wiener [16]. At the same time, we will be laying the groundwork for a more penetrating investigation of nonharmonic Gabor expansions in $L^2(\mathbb{R}^2)$.

Let $\{(\lambda_n, \mu_n)\}_{n\in\mathbb{Z}}$ be an arbitrary countable subset of $\mathbb{R}^2$ and

$$\{\varphi_n(\xi)\}_{n\in\mathbb{Z}} = \left\{M_{\mu_n} T_{\lambda_n} g(\xi)\right\}_{n\in\mathbb{Z}} = \left\{\sqrt{2}\, e^{2\pi i \mu_n \xi - \pi(\xi - \lambda_n)^2}\right\}_{n\in\mathbb{Z}}; \tag{2}$$

where $\xi \in \mathbb{R}^2$ or $\mathbb{C}$, be the corresponding Gabor system with respect to the Gaussian window g in $L^2(\mathbb{R}^2)$. If $\{\varphi_n\}_{n\in\mathbb{Z}}$ is incomplete in $L^2(\mathbb{R}^2)$ then the closed linear span $\mathcal{M}$ of $\{\varphi_n\}_{n\in\mathbb{Z}}$ is a proper subspace of $L^2(\mathbb{R}^2)$. By Hahn-Banach Theorem there exists a function F in $L^2(\mathbb{R}^2)$

such that $F|_{\mathcal{M}} = 0$ and $F \neq 0$. Riesz Representation Theorem implies that $F = F_\varphi$ for some φ in $L^2(\mathbb{R}^2)$ and

$$F(h) = F_\varphi(h) = \int_{\mathbb{R}^2} h\,\varphi\,d\xi; h \in L^2(\mathbb{R}^2).$$

For $(z, w) \in \mathbb{C}^2$ take

$$f(z, w) = \sqrt{2} \int_{\mathbb{R}^2} e^{2\pi i w\xi - \pi(\xi - z)^2}\,\varphi(\xi)\,d\xi, \tag{3}$$

then $f(\lambda_n, \mu_n) = F(\varphi_n) = 0$ (since $F|_{\mathcal{M}} = 0$).

Remark 3.1 The system (2) is incomplete in $L^2(\mathbb{R}^2)$ if and only if there exists a nontrivial entire function of the form (3) in the Gabor space $\mathcal{G}_g$, which is zero for every (λ_n, μ_n). Furthermore, since

$$f(z, w) = V_g \varphi(z, -w) = e^{i\pi z w} e^{-\pi \frac{|z|^2 + |w|^2}{2}} (B\varphi)(z, w),$$

we have

$$|f(z, w)| \leqq \|\varphi\|_2\, e^{\frac{\pi}{2}|(z,w)|^2}.$$

Theorem 3.2 Let $\{\lambda_n\}_{n \in \mathbb{Z}}$ be a symmetric sequence of real numbers ($\lambda_{-n} = -\lambda_n$). If the Gabor system

$$\left\{ \sqrt[4]{2}\, e^{2\pi i \lambda_n t - \pi(t - \lambda_n)^2} \right\}_{n \in \mathbb{Z}} \tag{4}$$

is exact in $L^2(\mathbb{R})$, then the product

$$\prod_{n=1}^{\infty} \left(1 - \frac{z^2}{\lambda_n^2} \right) e^{\frac{z^2}{\lambda_n^2}}$$

converges to an entire function which belongs to the Gabor space with Gaussian window in $L^2(\mathbb{R})$.

Proof. By Remark 3.1, if the system (4) is exact, then there exists an entire function $f(z)$ in the Gabor space $\mathcal{G}_g$ such that $f(\lambda_n) = 0$ for $n \neq 0$, and

$$f(z) = \sqrt[4]{2} \int_{\mathbb{R}} e^{2\pi i z t - \pi(t - z)^2}\,\varphi(t)\,dt; \quad \varphi \in L^2(\mathbb{R}).$$

Since $f(\lambda_n) = 0$ for $n \neq 0$ and the sequence $\{\lambda_n\}$ is symmetric, $\varphi(-t)$ has the same orthogonality properties as $\varphi(t)$. But by Remark 2.5, $\varphi(t)$ is unique, so $\varphi(t)$ must be even.

Hence $f(z)$ is even. Now $f(z)$ vanishes only at the λ_n with $n \neq 0$. Indeed, if $f(z)$ vanishes at $z = \gamma$, then the function

$$\tilde{f}(z) = \frac{zf(z)}{z - \gamma}$$

would also belong to $\mathcal{G}_g$ and would vanish at every λ_n. The system (4) would then be incomplete in $L^2(\mathbb{R})$, contrary to hypothesis.

Let us observe that the function $\tilde{f}$ belongs to $\mathcal{G}_g$. Since the Bargmann transform is related to the Gabor transform by the formula (1), it is sufficient to show that the function $e^{i\pi xy} e^{\frac{\pi}{2}(|x|^2 + |y|^2)} \tilde{f}(z)$; $z = x + iy$, belongs to the Fock space $\mathcal{F}(\mathbb{C})$. In other words, we must show that the integral

$$\int_{\mathbb{C}} \frac{|z|^2}{|z - \gamma|^2} \left| f(z) e^{i\pi xy} e^{\frac{\pi}{2}(|x|^2 + |y|^2)} \right|^2 e^{-\pi(|x|^2 + |y|^2)} dx\, dy; \quad z = x + iy,$$

is finite. Since $\lim_{z \to \infty} \left| \frac{z}{z - \gamma} \right| = 1$, we have $\left| \frac{z}{z - \gamma} \right| \leq 3/2$ outside a square T with complement T^c. Thus the above integral is no larger than

$$\int_T \frac{|z|^2}{|z - \gamma|^2} \left| f(z) e^{i\pi xy} e^{\frac{\pi}{2}(|x|^2 + |y|^2)} \right|^2 e^{-\pi(|x|^2 + |y|^2)} dx\, dy$$

$$+ 9/4 \int_{T^c} \left| f(z) e^{i\pi xy} e^{\frac{\pi}{2}(|x|^2 + |y|^2)} \right|^2 e^{-\pi(|x|^2 + |y|^2)} dx\, dy$$

$$\leqq \int_T \frac{|z|^2}{|z - \gamma|^2} \left| f(z) e^{i\pi xy} e^{\frac{\pi}{2}(|x|^2 + |y|^2)} \right|^2 e^{-\pi(|x|^2 + |y|^2)} dx\, dy$$

$$+ 9/4 \int_{\mathbb{C}} \left| f(z) e^{i\pi xy} e^{\frac{\pi}{2}(|x|^2 + |y|^2)} \right|^2 e^{-\pi(|x|^2 + |y|^2)} dx\, dy.$$

In the last expression, since T is compact the first integral is finite, and since the function $f(z) e^{i\pi xy} e^{\frac{\pi}{2}(|x|^2 + |y|^2)}$ is in the Fock space $\mathcal{F}(\mathbb{C})$, so is the second integral. Next since

$$|f(z)| \leqq \|\varphi\|_2\, e^{\frac{\pi}{2}|z|^2},$$

the order of growth of f is at most 2, and by Hadamard's factorization theorem,

$$f(z) = e^{Az} \prod_{n=1}^{\infty} \left(1 - \frac{z^2}{\lambda_n^2} \right) e^{\frac{z^2}{\lambda_n^2}}; \quad A \in \mathbb{R}.$$

Since $f(z)$ and the canonical product are both even, $A = 0$ and

$$f(z) = \prod_{n=1}^{\infty} \left(1 - \frac{z^2}{\lambda_n^2} \right) e^{\frac{z^2}{\lambda_n^2}}.$$

We have the following version of Plancherel-Pólya theorem. We give the proof which is similar to [11, Th. 2.16] for the sake of completeness.

Theorem 3.3 (Plancherel-Pólya). If $f(z)$ is an entire function of order of growth$\leq \tau$ and if for some positive number p,

$$\int_{-\infty}^{\infty} |f(x)|^p \, dx < \infty,$$

then

$$\int_{-\infty}^{\infty} |f(x+iy)|^p \, dx \leq e^{p\tau|y|} \int_{-\infty}^{\infty} |f(x)|^p \, dx.$$

The proof will require two preliminary lemmas. Suppose that $q(z)$ is a non constant continuous function in the closed upper half-plane, $\mathrm{Im}z \geq 0$, and analytic in its interior. Let a and p be positive real numbers and put

$$\mathcal{Q}(z) = \int_{-a}^{a} |q(z+t)|^p \, dt.$$

It is clear that $\mathcal{Q}(z)$ is continuous for $\mathrm{Im}z \geq 0$. Since $|q(z)|^p$ is subharmonic for $\mathrm{Im}z > 0$ (see [12, p.83]), so is $\mathcal{Q}(z)$.

Lemma 3.4 Let $q(z)$ be a function of order of growth$\leq \tau$ in the half-plane $\mathrm{Im}z \geq 0$ and suppose that the following quantities are both finite:

$$M = \sup_{-\infty < x < \infty} \mathcal{Q}(x) \quad and \quad N = \sup_{y > 0} \mathcal{Q}(iy).$$

Then on this half-plane,

$$\mathcal{Q}(z) \leq max(M, N).$$

Proof. Since $q(z)$ is of order of growth$\leq \tau$, then there exist positive numbers A and B such that

$$|q(z)| \leq Ae^{B|z|^\tau} \quad (\mathrm{Im}z \geq 0). \tag{5}$$

For each positive number ε, define the auxiliary function

$$q_\varepsilon(z) = q(z)e^{-\varepsilon(\lambda(z+a))^{3/2}}, \tag{6}$$

where $\lambda = e^{-i\pi/4}$. The exponent of e in (6) has two possible determinations in the half-plane $\text{Im} z > 0$; we choose the one whose real part is negative in the quarter-plane $x > -a$, $y \geq 0$. Put

$$Q_\varepsilon(z) = \int_{-a}^{a} |q_\varepsilon(z+t)|^p \, dt,$$

which is then defined and continuous in the upper half-plane $\text{Im} z \geq 0$, and subharmonic in its interior. A simple calculation involving (5) and (6) shows that in the quarter plane $x > -a$, $y \geq 0$,

$$|q_\varepsilon(z)| \leq A e^{B|z|^\tau - \varepsilon\gamma|z+a|^{3/2}}, \tag{7}$$

where $\gamma = \cos 3\pi/8$, and $|q_\varepsilon(z)| \leq |q(z)|$. Hence

$$Q_\varepsilon(z) \leq Q(z)(x \geq 0, y \geq 0),$$

and in particular

$$Q_\varepsilon(x) \leq M \quad \text{for} \quad x \geq 0 \quad \text{and} \quad Q_\varepsilon(iy) \leq N \quad \text{for} \quad y \geq 0.$$

Let z_0 be a fixed but arbitrary point in the first quadrant. We shall apply the maximum principle to $Q_\varepsilon(z)$ in the region $\Omega = \{z : \text{Re} z \geq 0, \text{Im} z \geq 0, |z| \leq R\}$, choosing R large enough so that (i) $z_0 \in \Omega$, and (ii) the maximum value of $Q_\varepsilon(z)$ on Ω is not attained on the circular $arc|z| = R$ (this is possible by virtue of (7)). Since $Q_\varepsilon(z)$ does not reduce to a constant, the maximum value of $Q_\varepsilon(z)$ on Ω must be attained on one of the coordinate axes, and in particular,

$$Q_\varepsilon(z_0) \leq max(M, N).$$

Now let $\varepsilon \to 0$. This establishes the result for the first quadrant; the proof for the second quadrant is similar.

Lemma 3.5 In addition to the hypotheses of Lemma 3.4, suppose that

$$\lim_{y \to \infty} q(x+iy) = 0 \tag{8}$$

uniformly in x, for $-a \leq x \leq a$. Then

$$Q(z) \leq M, \quad \text{Im} z \geq 0.$$

Proof. It is sufficient to show that $N \leq M$. By virtue of (8), we see that the function $Q(iy)$ tends to zero as $y \to \infty$, and so must attain its least upper bound N for some finite value of y, say $y = y_0$. If $y_0 = 0$, then

$$N = Q(iy_0) = Q(0) \leq M.$$

If $y_0 > 0$, then the maximum principle shows that the least upper bound of $Q(z)$ in the half-plane $\mathrm{Im}\, z \geq 0$ cannot be attained at the interior point $z = iy_0$. Therefore, by Lemma 3.4,

$$N = Q(iy_0) < max(M, N),$$

and again $N < M$.

Theorem 3.3 now follows.

Proof of Theorem 3.3. It is sufficient to prove the theorem when $y > 0$ and $f(z)$ is not identically zero. Let ε be a fixed positive number and consider the function

$$q(z) = f(z)e^{i(\tau+\varepsilon)z}.$$

It is easy to see that, for each positive number a, the functions $q(z)$ and $Q(z)$ satisfy the conditions Lemmas 3.4 and 3.5. Consequently, for $y > 0$,

$$Q(iy) \leq M < \int_{-\infty}^{\infty} |q(x)|^p dx.$$

This together with the definitions of $q(z)$ and $Q(z)$ implies

$$e^{-p(\tau+\varepsilon)y} \int_{-a}^{a} |f(x+iy)|^p dx < \int_{-\infty}^{\infty} |f(x)|^p dx.$$

To get the result, first let $a \to \infty$, then let $\varepsilon \to 0$.

Proposition 3.6 Let $f(z,w)$ be an entire function of order of growth $\leq \tau$ and suppose that $\{\lambda_n\}, \{\mu_n\}$ are increasing sequences of real numbers such that

$$\lambda_{n+1} - \lambda_n \geq \varepsilon_1 > 0 \quad \text{and} \quad \mu_{n+1} - \mu_n \geq \varepsilon_2 > 0.$$

If for some positive number p,

$$\sup_n \int_{-\infty}^{\infty} |f(x_z, \mu_n)|^p \, dx_z < \infty \quad \text{and} \quad \sup_n \int_{-\infty}^{\infty} |f(\lambda_n, x_w)|^p \, dx_w < \infty, \tag{9}$$

then

$$\sum_n |f(\lambda_n, \mu_n)|^p < \infty.$$

Proof. First, using the Plancherel-Pólya Theorem, observe that conditions (9) imply that

$$\sup_n \int_{-\infty}^{\infty} |f(z, \mu_n)|^p \, dx_z \leqq e^{p\tau|y_z|} \int_{-\infty}^{\infty} \sup_n |f(x_z, \mu_n)|^p \, dx_z,$$

and

$$\sup_n \int_{-\infty}^{\infty} |f(\lambda_n, w)|^p \, dx_w \leqq e^{p\tau|y_w|} \int_{-\infty}^{\infty} \sup_n |f(\lambda_n, x_w)|^p \, dx_w.$$

Now since $|f|^p$ is plurisubharmonic, the inequality

$$|f(z_0, w_0)|^p \leqq \frac{1}{2\pi} \int_0^{2\pi} |f((z_0, w_0) + (\zeta, \eta)e^{i\theta})|^p \, d\theta \qquad (10)$$

holds for all values of (ζ, η). Fix $\eta = 0$, multiply both sides of (10) by ζ and integrate between 0 and δ_1,

$$\int_0^{\delta_1} |f(z_0, w_0)|^p \zeta \, d\zeta \leqq \frac{1}{2\pi} \int_0^{\delta_1} \int_0^{2\pi} |f(z_0 + \zeta e^{i\theta}, w_0)|^p \, d\theta \, \zeta \, d\zeta.$$

Then

$$|f(z_0, w_0)|^p \leqq \frac{1}{\pi\delta_1^2} \iint_{\Omega_1} |f(z, w_0)|^p \, dx_z \, dy_z,$$

where $\Omega_1 = \{(z, w_0) : |z - z_0| \leqq \delta_1\}$. Similarly fix $\zeta = 0$, multiply both sides of (10) by η and integrate between 0 and δ_2,

$$|f(z_0, w_0)|^p \leqq \frac{1}{\pi\delta_2^2} \iint_{\Omega_2} |f(z_0, w)|^p \, dx_w \, dy_w,$$

where $\Omega_2 = \{(z_0, w) : |w - w_0| \leqq \delta_2\}$. Then

$$2|f(z_0, w_0)|^p \leqq \frac{1}{\pi\delta_1^2} \iint_{\Omega_1} |f(z, w_0)|^p \, dx_z \, dy_z$$

$$+ \frac{1}{\pi\delta_2^2} \iint_{\Omega_2} |f(z_0, w)|^p \, dx_w \, dy_w.$$

Now let $\Omega_1^n = \{(\lambda_n + z, \mu_n) : |z| \leq \delta_1\}$ and $\Omega_2^n = \{(\lambda_n, \mu_n + w) : |w| \leq \delta_2\}$, then

$$
\begin{aligned}
2 \sum_n |f(\lambda_n, \mu_n)|^p &\leq \sum_n \Big(\frac{1}{\pi \delta_1^2} \iint_{\Omega_1^n} |f(\lambda_n + z, \mu_n)|^p \, dx_z \, dy_z \\
&\quad + \frac{1}{\pi \delta_2^2} \iint_{\Omega_2^n} |f(\lambda_n, \mu_n + w)|^p \, dx_w \, dy_w \Big) \\
&\leq \sum_n \Big(\frac{1}{\pi \delta_1^2} \int_{-\delta_1}^{\delta_1} \int_{-\delta_1}^{\delta_1} |f(\lambda_n + z, \mu_n)|^p \, dx_z \, dy_z \\
&\quad + \frac{1}{\pi \delta_2^2} \int_{-\delta_2}^{\delta_2} \int_{-\delta_2}^{\delta_2} |f(\lambda_n, \mu_n + w)|^p \, dx_w \, dy_w \Big) \\
&= \sum_n \Big(\frac{1}{\pi \delta_1^2} \int_{-\delta_1}^{\delta_1} \int_{\lambda_n - \delta_1}^{\lambda_n + \delta_1} |f(z, \mu_n)|^p \, dx_z \, dy_z \\
&\quad + \frac{1}{\pi \delta_2^2} \int_{-\delta_2}^{\delta_2} \int_{\mu_n - \delta_2}^{\mu_n + \delta_2} |f(\lambda_n, w)|^p \, dx_w \, dy_w \Big).
\end{aligned}
$$

It is clear that the last expression above is no larger than

$$
\begin{aligned}
\sum_n \Big(\frac{1}{\pi \delta_1^2} &\int_{-\delta_1}^{\delta_1} \int_{\lambda_n - \delta_1}^{\lambda_n + \delta_1} \sup_n |f(z, \mu_n)|^p \, dx_z \, dy_z \\
&+ \frac{1}{\pi \delta_2^2} \int_{-\delta_2}^{\delta_2} \int_{\mu_n - \delta_2}^{\mu_n + \delta_2} \sup_n |f(\lambda_n, w)|^p \, dx_w \, dy_w \Big).
\end{aligned}
$$

Now for $\delta_1 = \frac{\varepsilon_1}{2}$ and $\delta_2 = \frac{\varepsilon_2}{2}$, the intervals $(\lambda_n - \delta_1, \lambda_n + \delta_1)$ are pairwise disjoint, and similarly for the intervals $(\mu_n - \delta_2, \mu_n + \delta_2)$, thus

$$
\begin{aligned}
2 \sum_n |f(\lambda_n, \mu_n)|^p &\leq \frac{1}{\pi \delta_1^2} \int_{-\delta_1}^{\delta_1} \int_{-\infty}^{\infty} \sup_n |f(z, \mu_n)|^p \, dx_z \, dy_z \\
&\quad + \frac{1}{\pi \delta_2^2} \int_{-\delta_2}^{\delta_2} \int_{-\infty}^{\infty} \sup_n |f(\lambda_n, w)|^p \, dx_w \, dy_w.
\end{aligned}
$$

We conclude that

$$
\begin{aligned}
2 \sum_n |f(\lambda_n, \mu_n)|^p &\leq \frac{1}{\pi \delta_1^2} \int_{-\delta_1}^{\delta_1} \Big(e^{p\tau |y_z|} \int_{-\infty}^{\infty} \sup_n |f(x_z, \mu_n)|^p \, dx_z \Big) dy_z \\
&\quad + \frac{1}{\pi \delta_2^2} \int_{-\delta_2}^{\delta_2} \Big(e^{p\tau |y_w|} \int_{-\infty}^{\infty} \sup_n |f(\lambda_n, x_w)|^p \, dx_w \Big) dy_w \\
&= B_1 \sup_n \int_{-\infty}^{\infty} |f(x_z, \mu_n)|^p \, dx_z \\
&\quad + B_2 \sup_n \int_{-\infty}^{\infty} |f(\lambda_n, x_w)|^p \, dx_w < \infty,
\end{aligned}
$$

where $B_1 = B_1(p, \tau, \varepsilon_1)$ and $B_2 = B_2(p, \tau, \varepsilon_2)$.

Remark 3.7 In the above proposition, if we replace the conditions (9) by

$$\int_{-\infty}^{\infty} \int_{-\infty}^{\infty} |f(x_z, x_w)|^p dx_z \, dx_w < \infty$$

the interior integral is finite everywhere except on a null set. If we use Fubini to change the order of integration, we get a null set for the integral against the second variable. If we know that none of λ_n's and μ_n's lie in these null sets, the conclusion still holds.

Theorem 3.8 If $\{\lambda_n\}_{n\in\mathbb{Z}}$ and $\{\mu_n\}_{n\in\mathbb{Z}}$ are separated sequences of real numbers such that $0 \leq \lambda_n \leq 1$ and $0 \leq \mu_n \leq 1$ for each n, then the Gabor system (2) forms a Bessel sequence in $L^2(\mathbb{R}^2)$. If $\sum_n |c_n|^2 < \infty$, then the Gabor expansion

$$\sum_n c_n e^{2\pi i \mu_n \xi - \pi(\xi - \lambda_n)^2}$$

converges in mean to an element of $L^2(\mathbb{R}^2)$.

Proof. If $\phi \in L^2(\mathbb{R}^2)$ then the inner product

$$a_n = \langle \sqrt{2} e^{2\pi i \mu_n \xi - \pi(\xi - \lambda_n)^2}, \phi(\xi) \rangle;$$

is just the value $f(\lambda_n, \mu_n)$ of the entire function

$$f(z, w) = \sqrt{2} \int_{\mathbb{R}^2} \varphi(\xi) \, e^{2\pi i w \xi - \pi(\xi - z)^2} \, d\xi; \, \varphi(\xi) = \overline{\phi(\xi)},$$

in the Gabor space $\mathcal{G}_g$ and f is of order of growth 2. we have

$$\sup_n \int_{-\infty}^{\infty} |f(x_z, \mu_n)|^p \, dx_z$$

$$\leqq 2^{p/2} M^p e^{\pi} \int_{-\infty}^{\infty} \left[\int_{\mathbb{R}} e^{-2\pi(x_\xi - x_z)^2} \, dx_\xi \int_{\mathbb{R}} e^{-2\pi(|y_\xi|-1)^2} \, dy_\xi \right]^{p/2} dx_z < \infty,$$

and similarly

$$\sup_n \int_{-\infty}^{\infty} |f(\lambda_n, x_w)|^p \, dx_w < \infty.$$

Therefore f satisfies conditions (9) and by Proposition 3.6 we have

$$\sum_n |\langle \sqrt{2}\, e^{2\pi i \mu_n \xi - \pi(\xi - \lambda_n)^2}, \phi(\xi) \rangle|^2 = \sum_n |a_n|^2 = \sum_n |f(\lambda_n, \mu_n)|^2 < \infty.$$

Thus the Gabor system (2) forms a Bessel sequence in $L^2(\mathbb{R}^2)$. The second part follows from the first by Proposition 2.4.

Paley and Wiener, showed in Theorem XLII of [16] that whenever

$$\lim_{n \to \pm\infty} (\lambda_{n+1} - \lambda_n) = \infty,$$

for a sequence of real numbers $\{\lambda_n\}$, then the exponentials are weakly independent over an arbitrarily short interval: $\sum a_n e^{i\lambda_n t} = 0$ only when all the a_n are zero. The next lemma states a similar statement for the set of complex exponentials replaced by the system (4). Here l.i.m. is used to show the limit in mean-square in L^2. The proof is almost identical to that of Paley and Wiener.

Lemma 3.9 Let no a_n vanish, $\sum_{-\infty}^{\infty} |a_n|^2$ converge, and let

$$\cdots < \lambda_{-n} < \cdots < \lambda_{-1} < \lambda_0 < \lambda_1 < \cdots < \lambda_n < \cdots$$

such that

$$\lim_{n \to \pm\infty} (\lambda_{n+1} - \lambda_n) = \infty.$$

Let

$$f(t) = \text{l.i.m.}_{N \to \infty} \sum_{-N}^{N} a_n\, e^{2\pi i \lambda_n t - \pi(t - \lambda_n)^2};$$

over every finite range. If $f(t)$ is equivalent to zero over any interval (a, b) then $f(t)$ is equivalent to zero over every interval, and all the a_n's vanish.

Now we want to show that if the separation of the λ_n's is great enough then system (4) is a Riesz-Fischer sequence.

Theorem 3.10 Let $\{\lambda_n\}$ be a sequence of real numbers whose differences are nondecreasing and satisfy

$$\sum \frac{1}{(\lambda_{k+1} - \lambda_k)^2} < \infty.$$

Then the Gabor system (4) is a Riesz-Fischer sequence in $L^2(\mathbb{R})$.

Proof. We adapt the proof of [16, Th. 1]. By the second part of the Proposition 2.4 we have to show that for all finite sequences of scalars $\{c_n\}$ and some constant $m > 0$,

$$m \sum |c_n|^2 \leqq \left\| \sum c_n \sqrt[4]{2}\, e^{2\pi i \lambda_n t - \pi(t-\lambda_n)^2} \right\|^2. \tag{11}$$

Using c to denote an l^2 sequence $\{c_1, c_2, \cdots\}$, inequality (11) is the same as

$$\frac{\langle Gc, c \rangle_{l^2}}{\langle c, c \rangle_{l^2}} \geqq m,$$

where the l^2 operator G is the Gram matrix of the members of the Gabor system (4). It is to be shown that the eigenvalues of finite subsections of G are bounded away from zero, which in turn follows from these two conditions:

(1) $Gv = 0$ implies $v = 0$, for every l^2 sequence v.

(2) $G = I + M$, where M is a compact operator.

The first condition is satisfied by Lemma 3.9. To verify condition (2), observe that the entries of $G = I + M$ are

$$g_{nm} = \sqrt{2} \int_{-\infty}^{\infty} e^{2\pi i(\lambda_n - \lambda_m)t - \pi(t-\lambda_n)^2 - \pi(t-\lambda_m)^2}\, dt.$$

Now M can be shown to be compact by showing that its Schmidt norm is finite. Since G is symmetric, it suffices to show that

$$\sum_{n=1}^{\infty} \sum_{m=n+1}^{\infty} g_{nm}^2 < \infty.$$

The sum is bounded above,

$$\sum_{n=1}^{\infty} \sum_{m=n+1}^{\infty} g_{nm}^2 = 2 \sum_{n=1}^{\infty} \sum_{m=n+1}^{\infty} \left(\int_{-\infty}^{\infty} e^{2\pi i(\lambda_n - \lambda_m)t - \pi(t-\lambda_n)^2 - \pi(t-\lambda_m)^2}\, dt \right)^2$$

$$\leqq 2 \sum_{n=1}^{\infty} \sum_{m=n+1}^{\infty} \left(\int_{-\infty}^{\infty} e^{2\pi i(\lambda_n - \lambda_m)t}\, dt \right)^2$$

$$= 2 \sum_{n=1}^{\infty} \sum_{m=n+1}^{\infty} \lim_{A \to \infty} \left(\int_{-A}^{A} e^{2\pi i(\lambda_n - \lambda_m)t}\, dt \right)^2$$

$$\leqq \frac{2}{\pi^2} \sum_{n=1}^{\infty} \sum_{m=n+1}^{\infty} \frac{1}{(\lambda_n - \lambda_m)^2}$$

$$< \frac{2}{\pi^2} \sum_{n=1}^{\infty} \sum_{m=n+1}^{\infty} \frac{1}{(\lambda_{n+1} - \lambda_n)^2 (m-n)^2},$$

where $(\lambda_m - \lambda_n) \leqq (\lambda_{n+1} - \lambda_n)(m+n)$ follows from the assumption that differences are nondecreasing. Letting $k = m + n$, one concludes that

$$\sum_{n=1}^{\infty} \sum_{m=n+1}^{\infty} g_{nm}^2 < C \sum_{n=1}^{\infty} \frac{1}{(\lambda_{n+1} - \lambda_n)^2} < \infty,$$

establishing the theorem.

Theorem 3.11 Let

$$f(z) = \int_{-\infty}^{\infty} \alpha(t) e^{2\pi i z t - \pi(t-z)^2} \, dt,$$

where $\alpha \in L^2(\mathbb{R})$. If $f(\mu) = 0$ and $g(z) = \dfrac{z - \lambda}{z - \mu} f(z)$, then there exists a function β in $L^2(\mathbb{R})$ such that

$$g(z) = \int_{-\infty}^{\infty} \beta(t) e^{2\pi i z t - \pi(t-z)^2} \, dt. \tag{12}$$

Moreover,

$$\beta(t) = \alpha(t) + 2\pi(i+1)(\lambda - \mu) e^{-2\pi i \mu t + \pi(t-\mu)^2} \int_{-\infty}^{t} \alpha(s) e^{2\pi i \mu s - \pi(s-\mu)^2} \, ds \tag{13}$$

almost everywhere on $\mathbb{R}$.

Proof. To motivate the proof, let us suppose that $g(z)$ is in fact representable in the form (12), and try to deduce (13). If (12) holds, then

$$\frac{1}{z - \mu} \int_{-\infty}^{\infty} \alpha(t) e^{2\pi i z t - \pi(t-z)^2} \, dt = \frac{1}{z - \lambda} \int_{-\infty}^{\infty} \beta(t) e^{2\pi i z t - \pi(t-z)^2} \, dt.$$

The trick in solving for $\beta(t)$ is to transform each of these integrals by first rewriting $e^{2\pi i z t - \pi(t-z)^2}$ as

$$e^{2\pi i z t - \pi(t-z)^2} = e^{2\pi i (z-\mu)t + 2\pi i \mu t - \pi(t-\mu)^2 + \pi(z-\mu)(2t-z-\mu)}.$$

and then integrating by parts. When this is done, the result is

$$\frac{1}{z - \mu} \int_{-\infty}^{\infty} \alpha(t) e^{2\pi i z t - \pi(t-z)^2} \, dt = \int_{-\infty}^{\infty} \alpha_1(t) e^{2\pi i z t - \pi(t-z)^2} \, dt,$$

with

$$\alpha_1(t) = -2(i+1)\pi e^{-2\pi i \mu t + \pi(t-\mu)^2} \int_{-\infty}^t \alpha(s) e^{2\pi i \mu s - \pi(s-\mu)^2} \, ds,$$

and

$$\frac{1}{z-\lambda} \int_{-\infty}^{\infty} \beta(t) e^{2\pi i z t - \pi(t-z)^2} \, dt = \int_{-\infty}^{\infty} \beta_1(t) e^{2\pi i z t - \pi(t-z)^2} \, dt,$$

with

$$\beta_1(t) = -2(i+1)\pi e^{-2\pi i \lambda t + \pi(t-\lambda)^2} \int_{-\infty}^t \beta(s) e^{2\pi i \lambda s - \pi(s-\lambda)^2} \, ds.$$

It follows that $\alpha_1(t) = \beta_1(t)$ almost everywhere on $\mathbb{R}$, and so

$$e^{2\pi i(\lambda-\mu)t + \pi(\lambda-\mu)(2t-(\lambda+\mu))} \int_{-\infty}^t \alpha(s) e^{2\pi i \mu s - \pi(s-\mu)^2} \, ds$$

$$= \int_{-\infty}^t \beta(s) e^{2\pi i \lambda s - \pi(s-\lambda)^2} \, ds.$$

To obtain (13), differentiate both sides of this equation with respect to t. Now simply observe that all of the above steps are reversible, that is $\beta \in L^2(\mathbb{R})$.

Remark 3.12 A similar result holds when f is of the form

$$f(z) = \int_{-\infty}^{\infty} e^{2\pi i z t - \pi(t-z)^2} \, d\alpha(t),$$

and α is of bounded variation on $\mathbb{R}$, only now

$$g(z) = \int_{-\infty}^{\infty} e^{2\pi i z t - \pi(t-z)^2} \, d\beta(t),$$

with

$$d\beta(t) = d\alpha(t) + 2\pi i(\lambda - \mu) e^{-2\pi i \mu t + \pi(t-\mu)^2} \int_{-\infty}^t e^{2\pi i \mu s - \pi(s-\mu)^2} \, d\alpha(s).$$

Corollary 3.13 The completeness of system (4) is unaffected if one λ_n is replaced by another number.

Nowak [18] showed that the deficit of the regular Gabor system generated by $h \in L^2(\mathbb{R}^d)$ and $a, b > 0$ is either zero or infinite if the system is a Bessel sequence in $L^2(\mathbb{R}^d)$. The next result on the deficit of the irregular Gabor system (4) is proved as in [19, Th. 4.6]. Here we give the proof for the sake of completeness.

Theorem 3.14 If $\{\lambda_n\}$ is a separated sequence of real numbers such that

$$\lambda_{n+1} - \lambda_n > 1; \; (n = 0, \pm 1, \pm 2, \cdots)$$

then the Gabor system (4) has infinite deficiency in $L^2(\mathbb{R})$.

Proof. Let N be a fixed but arbitrary positive integer. If K is large enough, then we can replace

$$\lambda_0, \lambda_1, \cdots, \lambda_K$$

by

$$\mu_0, \mu_1, \cdots, \mu_{K+N+1},$$

so that the resulting sequence, relabeled $\{\mu_n\}$, satisfies

$$\inf_n(\mu_{n+1} - \mu_n) > 1.$$

By Theorem 3.10 there is a function $\varphi \in L^2(\mathbb{R})$ such that

$$\int_{\mathbb{R}} \varphi(t) \sqrt[4]{2}\, e^{-2\pi i \mu_n t - \pi(t - \mu_n)^2}\, dt = \begin{cases} 1 \text{ if } n = 0, \\ 0 \text{ if } n \neq 0 \end{cases}.$$

Thus the system

$$\left\{ \sqrt[4]{2}\, e^{2\pi i \mu_n t - \pi(t - \mu_n)^2} : n \neq 0 \right\}$$

is incomplete in $L^2(\mathbb{R})$, and we conclude by the above corollary that the deficiency of the system in $L^2(\mathbb{R})$ is at least N.

4. Stability

In this section we study stability of sampling sets in Gabor spaces. Here we let $\mathcal{G}_h$ to be the Gabor space of a Gabor window $h \in L^2(\mathbb{R}^d)$, and Λ be a discrete set in $\mathbb{C}^d$.

Proposition 4.1 [17, Cor. 3.2.3] (Inversion formula for the Gabor transform) Let $h, \gamma \in L^2(\mathbb{R}^d)$ be such that $\langle h, \gamma \rangle \neq 0$, and we consider $V_h f(z) = \langle f, h_z \rangle$, for every $f \in L^2(\mathbb{R}^d)$. Then it is fulfilled that $V_h f \in L^2(\mathbb{C}^d)$. Moreover we have inversion formula given by:

$$f(t) = \frac{1}{\langle h, \gamma \rangle} \int_{\mathbb{C}^d} V_h f(z)\, M_y T_x \gamma(t)\, dy\, dx; \quad z = x + iy \in \mathbb{C}^d.$$

The image of $L^2(\mathbb{R}^d)$ under the Gabor transform with the window h, forms a reproducing kernel Hilbert space

$$\mathcal{G}_h = \{V_h f : f \in L^2(\mathbb{R}^d)\}$$

(a closed subspace of $L^2(\mathbb{C}^d)$) which is called *Gabor space* or *model space*.

The following result is proved for $d = 1$ in [19, Prop. 1.29], the proof given here is based on [17].

Proposition 4.2 The Gabor space $\mathcal{G}_h$ of a Gabor window $h \in L^2(\mathbb{R}^d)$ is a Hilbert subspace of $L^2(\mathbb{C}^d)$ that is characterized for the following reproducing kernel:

$$k_h(z, z_0) = k(z, z_0) = k_{z_0}(z) = \langle h_{z_0}, h_z \rangle.$$

That is, $F \in \mathcal{G}_h$ if and only if

$$F \in L^2(\mathbb{C}^d)$$

and

$$F(z_0) = \int_{\mathbb{C}^d} F(z) \overline{k(z, z_0)} \, dx \, dy. \tag{14}$$

Proof. We introduced the inversion formula for the Gabor transform in Proposition 4.1. Without loss of generality, we may assume that $\|h\| = 1$. Now for $z_0 = x_0 + iy_0 \in \mathbb{C}^d$ we have

$$V_h f(z_0) = \int_{\mathbb{R}^d} f(t) \overline{M_{y_0} T_{x_0} h(t)} \, dt$$

$$= \int_{\mathbb{R}^d} \left(\int_{\mathbb{C}^d} V_h f(z) M_y T_x h(t) dx \, dy \right) \overline{M_{y_0} T_{x_0} h(t)} \, dt$$

where $z = x + iy$. Switching the integrals we have

$$V_h f(z_0) = \int_{\mathbb{C}^d} V_h f(z) \left(\int_{\mathbb{R}^d} M_y T_x h(t) \overline{M_{y_0} T_{x_0} h(t)} \, dt \right) dx \, dy$$

$$= \int_{\mathbb{C}^d} V_h f(z) \left(\int_{\mathbb{R}^d} h_z(t) \overline{h_{z_0}(t)} \, dt \right) dx \, dy,$$

that is,

$$F(z_0) = \int_{\mathbb{C}^d} F(z) \langle h_z, h_{z_0} \rangle dx \, dy$$

$$= \int_{\mathbb{C}^d} F(z) \overline{k(z, z_0)} dx \, dy.$$

Therefore k replays all the functions of the space, and as it belongs to this space, it is its reproducing kernel.

For $x, y \in \mathbb{R}^d$ we recall T_x describes a translation by x also called a time shift and M_y a modulation by y also called a frequency shift. So the operators of the form $M_y T_x$ or $T_x M_y$ are known as time-frequency shifts. They satisfy the commutation relations

$$T_x M_y f(t) = (M_y f)(t - x)$$
$$= e^{2\pi i y.(t-x)} f(t - x)$$
$$= e^{-2\pi i y.x} M_y T_x f(t)$$

Then we have

$$\langle h_z, h_{z_0} \rangle = \langle M_y T_x h, M_{y_0} T_{x_0} h \rangle$$
$$= \langle h, T_{-x} M_{y_0 - y} T_{x_0} h \rangle$$
$$= \langle h, e^{2\pi i x.(y_0 - y)} M_{y_0 - y} T_{x_0 - x} h \rangle$$
$$= e^{2\pi i x.(y - y_0)} \langle h, M_{y_0 - y} T_{x_0 - x} h \rangle$$
$$= e^{2\pi i x.(y - y_0)} \langle h, h_{z_0 - z} \rangle$$

In terms of $k_h(z) = k_h(z, 0) = \langle h, h_z \rangle$ one has

$$k_h(z, z_0) = \overline{\langle h_z, h_{z_0} \rangle} = \overline{e^{2\pi i x.(y - y_0)} \langle h, h_{z_0 - z} \rangle}$$
$$= \overline{e^{2\pi i x.(y - y_0)} k_h(z_0 - z)}$$

and hence the reproduction formula (14) takes the form

$$F(z_0) = \int_{\mathbb{C}^d} F(z) e^{2\pi i x.(y - y_0)} k_h(z_0 - z) dx \, dy$$

Using this notations we can deduce that

$$V_h f_{z_0}(z) = \langle f_{z_0}, h_z \rangle = e^{2\pi i x_0.(y_0 - y)} \langle f, h_{z - z_0} \rangle$$
$$= e^{2\pi i x_0.(y_0 - y)} V_h f(z - z_0).$$

In this way, to be consistent with the notation and the definition of the transform, we have to define the translations in $\mathbb{C}^d$ of a function $F \in \mathcal{G}_h$ (or in $L^2(\mathbb{C}^d)$ in a general way) as:

$$F_{z_0}(z) = e^{2\pi i x_0.(y_0 - y)} F(z - z_0).$$

It is necessary to observe that these translations do not coincide in general with the usual translation of $\mathbb{C}^d$. But if we look at the function, then we have

$$|F_{z_0}(z)| = |F(z - z_0)|.$$

Since in general the function $F(z - z_0)$ can not belong to $\mathcal{G}_h$. Taking this into account we can write the reproduction formula in a bit more compact way

$$F(z_0) = \int_{\mathbb{C}^d} F(z)k_z(z_0)dx\,dy$$

The Gabor space has certain good continuity properties. More precisely, the functions of the space will be uniformly continuous. For $F \in \mathcal{G}_h$, since F is defined as a definite integral, it is uniformly continuous with respect to the free variable of the integrand, i.e. for each $\varepsilon > 0$ there exists $\delta > 0$ such that if $|z_1 - z_2| < \delta$ by using triangle inequality we have

$$\big||F(z_1)| - |F(z_2)|\big| \leq |F(z_1) - F(z_2)| < \varepsilon$$

Ascensi [19] formalized this idea in the case $d = 1$ with the following result.

Proposition 4.3 [19, Prop. 5.1] Let $\mathcal{G}_h$ be the Gabor space of normalized Gabor window $h \in L^2(\mathbb{R}^d)$. Then, given ε there exists δ such that if $|z_1 - z_2| < \delta$, then for every $F \in \mathcal{G}_h$

$$F(z) = \langle f, h_z \rangle$$

it is fulfilled that

$$\big||F(z_1)| - |F(z_2)|\big| < \|F\|\varepsilon = \|f\|\varepsilon.$$

A good description of the Gabor space is most convenient if there are some complete characterizations. The best situation occurs when for some analyzing function the Gabor space is a space of holomorphic functions. The most important example and the only possible one is the Gaussian function, for which the Gabor space can be identified with the Fock space, in which the sampling and interpolation sets are completely characterized [20]. The following assertion is proved for $d = 1$ in [7], we do not know if the same holds in higher dimensions.

Problem 4.4 Consider the Gabor space with a Gabor window $h \in L^2(\mathbb{R}^d)$

$$\mathcal{G}_h = \left\{ F(z) = \int_{\mathbb{R}^d} f(t)e^{-2\pi i y \cdot t}\overline{h(t - x)}\,dt,\ f \in L^2(\mathbb{R}^d) \right\}.$$

Then this space is a space of antiholomorphic functions (i.e., $F(x, -y)$ is holomorphic), modulo a multiplication by a weight, if and only if h is a time-frequency translation of the Gaussian function.

As the space in that we will work is formed by continuous functions and it has reproducing kernel, we can sample the functions at any point. Given a set of points of $\mathbb{C}^d$ we can consider, for each $F \in \mathcal{G}_h$, the succession of values that F takes in this set.

Definition 4.5 A discrete set $\Lambda = \{z_j\}_{j\in\mathbb{Z}}$ in $\mathbb{C}^d$ is said to be a *sampling set* for $\mathcal{G}_h$ if there are constants $A, B > 0$ such that

$$A\|F\|^2 \leqq \sum_{j\in\mathbb{Z}} |F(z_j)|^2 \leqq B\|F\|^2 F \in \mathcal{G}_h.$$

These sets are very important since they correspond with frames. We give some properties of Gabor space and sampling set in the case $d > 1$. The case $d = 1$ was considered by Ascensi and Bruna in [7]. The proofs are essentially the same (with little changes in certain cases). Recall that $\mathcal{G}_h = \{V_h f : f \in L^2(\mathbb{R}^d)\}$ is the Gabor space. Here we assume that $\|h\| = 1$.

Proposition 4.6 If $\Lambda = \{z_j\}_{j\in\mathbb{N}}$ is a sampling set for $\mathcal{G}_h$ then Λ is a relatively separated set.

Proof. The proof is exactly similar to the proof of [7, Prop. 3.1].

Definition 4.7 Given a continuous function F defined in $\mathbb{C}^d$ we define its *local maximal function* as:

$$MF(z) = \sup_{|w-z|<1} |F(w)|$$

Lemma 4.8 Let $\Lambda = \{\lambda_j\}_{j\in\mathbb{Z}}$ be a separated set with separation constant ε. Then

$$\sum_{\lambda\in\Lambda} |k(\lambda)| < \frac{1}{c\varepsilon^{2d}}\|Mk\|_1,$$

where $c = m(B(0,1))$.

Proof. We suppose without loss of generality that $1 < \varepsilon < 2$. Then using sub-mean-value inequality

$$\sum_{\lambda\in\Lambda} |k(\lambda)| \leqq \sum_{\lambda\in\Lambda} \frac{1}{|B(\lambda,\varepsilon)|} \int_{B(\lambda,\varepsilon)} |k(z)|\, dm(z)$$

$$\leqq \sum_{\lambda\in\Lambda} \frac{1}{|B(\lambda,\varepsilon)|} \int_{B(\lambda,\varepsilon)} Mk(z)\, dm(z)$$

$$\leqq \sum_{\lambda\in\Lambda} \frac{1}{|B(\lambda,\varepsilon)|} \int_{\mathbb{C}^d} Mk(z)\, dm(z).$$

Since by hypothesis those balls are disjoint, then

$$\sum_{\lambda \in \Lambda} |k(\lambda)| < \frac{1}{c\varepsilon^{2d}} \|Mk\|_1.$$

.

Proposition 4.9 If Λ is a separated set, there exists $B > 0$ such that

$$\sum_{\lambda \in \Lambda} |F(\lambda)|^2 \leqq B\|F\|^2; \quad F \in \mathcal{G}_h.$$

Proof. Calculating directly we have that

$$
\begin{aligned}
\sum_{\lambda \in \Lambda} |F(\lambda)|^2 &= \sum_{\lambda \in \Lambda} \left| \int_{\mathbb{C}^d} F(z)\overline{k_\lambda(z)}\, dm(z) \right|^2 \\
&\leqq \sum_{\lambda \in \Lambda} \left(\int_{\mathbb{C}^d} |F(z)|^2 |k_\lambda(z)|\, dm(z) \right) \times \left(\int_{\mathbb{C}^d} |k_\lambda(z)|\, dm(z) \right) \\
&= \int_{\mathbb{C}^d} |F(z)|^2 \sum_{\lambda \in \Lambda} |k(z-\lambda)|\, dm(z) \times \int_{\mathbb{C}^d} |k(z)|\, dm(z).
\end{aligned}
$$

Here $\int_{\mathbb{C}^d} |k(z)|\, dm(z) = m < \infty$ because the kernel is integrable and also

$$\sum_{\lambda \in \Lambda} |k(z-\lambda)| = \sum_{\gamma \in (z-\Lambda)} |k(\gamma)|$$

is bounded independently of z, and since $z - \Lambda$ has the same separation constant as Λ we can apply Lemma 4.8. Then

$$
\begin{aligned}
\sum_{\lambda \in \Lambda} |F(\lambda)|^2 &\leqq \int_{\mathbb{C}^d} |F(z)|^2 \sum_{\gamma \in (z-\Lambda)} |k(\gamma)|\, dm(z) \int_{\mathbb{C}^d} |k(z)|\, dm(z) \\
&\leqq B\|F\|^2,
\end{aligned}
$$

where $B = \dfrac{m}{c\varepsilon^{2d}}\|Mk\|_1$ and ε is the separation constant of Λ.

Next, we want to know when the Gabor system $G(h, \Lambda)$ is a frame for $L^2(\mathbb{R}^d)$. First we observe that $V_h f(z) = \langle f, h_z \rangle$ and $\|V_h f\| = \|f\|$ (if $\|h\| = 1$). As we have bijective correspondence between $\mathcal{G}_h$ and $L^2(\mathbb{R}^d)$ by the Gabor transform, we can write

$$\sum_{\lambda \in \Lambda} |F(\lambda)|^2 = \sum_{\lambda \in \Lambda} |V_h f(\lambda)|^2 = \sum_{\lambda \in \Lambda} |\langle f, h_\lambda \rangle|^2$$

for every $f \in L^2(\mathbb{R}^d)$ or for every $V_h f \in \mathcal{G}_h$. Therefore the frame condition and that of sampling set are equivalent. We conclude that: given a discrete set $\Lambda \subset \mathbb{C}^d$ and a Gabor window $h \in L^2(\mathbb{R}^d)$, $G(h, \Lambda)$ is a frame for $L^2(\mathbb{R}^d)$ if and only if Λ is a sampling set for $\mathcal{G}_h$.

Let $\Sigma_\alpha = \{h \in C^\infty(\mathbb{R}) \cap L^2(\mathbb{R}) : h' + zh \in L^2(\mathbb{R}) \text{ and } \|h' + zh\|_2 \leq \alpha\|h\|_2 \text{ for } z(t) = t\, t \in \mathbb{R}\}$. It is clear that $C_c^\infty(\mathbb{R}) \subseteq \Sigma_\alpha$ and so Σ_α is a non empty set.

Lemma 4.10 Let $h \in \Sigma_\alpha$ and let $\{\lambda_n\}_{n\in\mathbb{N}}$ and $\{\mu_n\}_{n\in\mathbb{N}}$ be sequences of scalars and suppose that there exist positive numbers B and L such that

$$\sum_n |F(\lambda_n)|^2 \leq B\|F\|^2; (F \in \mathcal{G}_h)$$

and

$$|\mu_n - \lambda_n| \leq L, (n = 1, 2, 3, \cdots).$$

Then for every $F \in \mathcal{G}_h$

$$\sum_n |F(\lambda_n) - F(\mu_n)|^2 \leq B(e^{\alpha L} - 1)^2 \|F\|^2.$$

Proof. Let F be an element of $\mathcal{G}_h$. By expanding F in a Taylor series about λ_n, we find that

$$F(\mu_n) - F(\lambda_n) = \sum_{k=1}^\infty \frac{F^{(k)}(\lambda_n)}{k!} (\mu_n - \lambda_n)^k (n = 1, 2, 3, \cdots).$$

If ρ is an arbitrary positive number, then by multiplying and dividing the summand by ρ^k we find also that

$$|F(\mu_n) - F(\lambda_n)|^2 \leq \sum_{k=1}^\infty \frac{|F^{(k)}(\lambda_n)|^2}{\rho^{2k}k!} \sum_{k=1}^\infty \frac{\rho^{2k}|\mu_n - \lambda_n|^{2k}}{k!}$$

Since $\mathcal{G}_h$ is closed under differentiation and $\|F'\|_2 \leq \alpha\|F\|_2$ it follows that

$$\sum_n |F^{(k)}(\lambda_n)|^2 \leq B\|F^{(k)}\|^2$$

$$\leq B\alpha^{2k}\|F\|^2, \quad k = 1, 2, 3, \cdots$$

Therefore we obtain

$$\sum |F(\lambda_n) - F(\mu_n)|^2 \leq B\|F\|^2 \sum_{k=1}^\infty \frac{\alpha^{2k}}{\rho^{2k}k!} \sum_{k=1}^\infty \frac{(\rho L)^{2k}}{k!}$$

$$= B\|F\|^2 (e^{\frac{\alpha^2}{\rho^2}} - 1)(e^{\rho^2 L^2} - 1)$$

since $|\mu_n - \lambda_n| \leqq L$.

Now by choosing $\rho = \sqrt{\dfrac{\alpha}{L}}$ we get

$$\sum_n |F(\lambda_n) - F(\mu_n)|^2 \leqq B(e^{\alpha L} - 1)^2 \|F\|^2.$$

Theorem 4.11 If $\{\lambda_n\}_{n \in \mathbb{N}}$ is a sampling set for $\mathcal{G}_h (h \in \Sigma_\alpha)$ then there exists positive constant L such that if $\{\mu_n\}_{n \in \mathbb{N}}$ satisfies $|\lambda_n - \mu_n| \leq L$ for all n, then $\{\mu_n\}_{n \in \mathbb{N}}$ is also sampling set.

Proof. Since $\{\lambda_n\}_{n \in \mathbb{N}}$ is a sampling set for $\mathcal{G}_h$, then there exist positive constants A and B such that

$$A\|F\|^2 \leqq \sum_n |F(\lambda_n)|^2 \leqq B\|F\|^2$$

for every function F belonging to the Gabor space $\mathcal{G}_h$. Let $\{\mu_n\}_{n \in \mathbb{N}}$ be complex scalars for which $|\lambda_n - \mu_n| \leqq L (n = 1, 2, 3, \cdots)$. It is to be shown that if L is sufficiently small, then similar inequalities hold for the μ_n's.

By virtue of the previous lemma, for every $F \in \mathcal{G}_h$,

$$\sum_n |F(\lambda_n) - F(\mu_n)|^2 \leqq B(e^{\alpha L} - 1)^2 \|F\|^2$$

and since $\|F\|^2 \leqq \dfrac{1}{A} \sum_n |F(\lambda_n)|^2$ to have

$$\sum_n |F(\lambda_n) - F(\mu_n)|^2 \leqq C \sum_n |F(\lambda_n)|^2$$

where $C = \dfrac{B}{A}(e^{\alpha L} - 1)^2$. Applying Minkowski's inequality, we find that

$$\left| \sqrt{\sum |F(\lambda_n)|^2} - \sqrt{\sum |F(\mu_n)|^2} \right| \leqq \sqrt{C \sum |F(\lambda_n)|^2},$$

and hence

$$\sqrt{A}(1 - \sqrt{C})\|F\| \leqq \sqrt{\sum |F(\mu_n)|^2} \leqq \sqrt{B}(1 + \sqrt{C})\|F\|$$

for every F. Since C is less than 1 if L is sufficiently small, $\{\mu_n\}_{n\in\mathbb{N}}$ is a sampling set for $\mathcal{G}_h$.

Problem 4.12 Let $h, k \in L^2(\mathbb{R})$. It would be desirable to show that there exists $\varepsilon > 0$ such that if $\|k - h\| < \varepsilon$ and $\{\lambda_n\}$ is a sampling set for $\mathcal{G}_h$ then $\{\lambda_n\}$ is a sampling set for $\mathcal{G}_k$. If one can show this and $h \in \Sigma_\alpha$, then for each $k \in L^2(\mathbb{R})$ with $\|k - h\| < \varepsilon$ the stability result of Theorem 4.11 holds for $\mathcal{G}_k$ as well. Now since Σ_α is norm dense in $L^2(\mathbb{R})$, one could conclude that Theorem 4.11 holds for each $h \in L^2(\mathbb{R})$. At present we are not able to prove that a "small" perturbation does not effect sampling set.

Acknowledgement

This is part of the Ph.D. thesis of the first author in Tarbiat Modares University. She would like to thank Tarbiat Modares University and MSRT for moral and financial support. Part of this work appeared in a joint paper of the authors in Journal of Sciences, Islamic Republic of Iran.

Author details

Nafya Hameed Mohammad[1] and Massoud Amini[2]

*Address all correspondence to: nafya.mohammad@su.edu.iq; mamini@modares.ac.ir

1 Department of Mathematics, College of Education, Salahaddin University, Erbil, Iraq

2 Department of Mathematics, Faculty of Mathematical Sciences, Tarbiat Modares University, Tehran, Iran

References

[1] Gabor, D., Theory of Communication, *J. Inst. Elec. Eng.* 93 (1946) 429-457.

[2] Folland, G. B., *Harmonic Analysis in Phase Space*, Princeton University Press, Princeton, 1989.

[3] Feichtinger, H. G. and Gröchenig, K., Banach spaces related to integrable group representations and their atomic decompositions. I., *J. Funct. Anal.* 86(2) (1989) 307-340.

[4] Liang, T. and Juan, G., Fast parallel algorithms for discrete Gabor expansion and transform based on multirate filtering, *Science China* 55(2) (2012) 293-300.

[5] Wexler, J. and Raz, S., Discrete Gabor expansions, *Signal Processing* 21 (1990) 207-220.

[6] Borichev, A., Gröchenig, K. and Lyubarskii, Yu., Frame constants of Gabor frames near the critical density, *J. Math. Pures Appl.* 94 (2010) 170-182.

[7] Ascensi, G. and Bruna, J., Model space results for the Gabor and wavelet transforms, *IEEE Trans. Inform. Theory* 55(5) (2009) 2250-2259.

[8] Abreu, L. D., On the structure of Gabor and super Gabor spaces, *Monatsh Math.* 161 (2010) 237-253.

[9] Christensen, O., *Frames and Bases, An Introductory Course*, Birkhäuser, Boston, 2008.

[10] Duffin, R. J. and Schaeffer, A. C., A class of nonharmonic Fourier series, *Trans. Amer. Math. Soc.* 72(2) (1952) 341-366.

[11] Young, R. M., *An Introduction to Nonharmonic Fourier Series*, 2nd. ed., Academic Press, New York, 2001.

[12] Reid, R. M., A class of Riesz-Fischer sequences, *Proc. Amer. Math. Soc.* 123(3) (1995) 827-829.

[13] Jaffard, S., Pointwise and directional regularity of nonharmonic Fourier series, *Appl. Comput. Harmon. Anal.* 22(3) (2010) 251-266.

[14] Krantz, S. G., *Function Theory of Several Complex Variables*, 2nd ed., Amer. Math. Soc., Providence, 2001.

[15] Shabat, B. V., *Introduction to Complex Analysis, Part II, Functions of Several Variables*, Amer. Math. Soc., Providence, 1992.

[16] Paley, R. C. and Wiener, N., *Fourier Transforms in the Complex Domain*, Colloquium Publications, vol. 19, Amer. Math. Soc., New York, 1934.

[17] Gröchenig, K., *Foundations of Time-Frequency Analysis*, Birkhäuser, Boston, 2001.

[18] Nowak, K., *Excess of Gabor Frames*, University of Wien, Wien, 2006.

[19] Ascensi, G., *Generators of $L^p(\mathbb{R})$ by translations in time and frequency*, Ph.D. Thesis, Univ. of Wien, 2007.

[20] Seip, K., Density theorems for sampling and interpolation in the Bargmann-Fock space I, *J. Reine Angew. Math.* 429 (1992) 91-106.

4

Experimental Data Deconvolution Based on Fourier Transform Applied in Nanomaterial Structure

Adrian Bot, Nicolae Aldea and Florica Matei

Additional information is available at the end of the chapter

1. Introduction

In many kinds of experimental measurements, such as astrophysics, atomic physics, biophysics, geophysics, high energy physics, nuclear physics, plasma physics, solid state physics, bending or torsion elastic, heat propagation or statistical mechanics, the signal measured in the laboratory can be expressed mathematically as a convolution of two functions. The first represents the resolution function called the instrumental signal, which is specific for each setup, and the second is the true sample that contains all physical information. These phenomena can be modelled by an integral equation, which means the unknown function is under the integral operator. The most important type of integral equation applied in physical and technical signal treatments is the Fredholm integral equation of the first kind. The opposite process when used for true sample function determination is known in the literature as experimental data deconvolution. Solution determination of the deconvolution equation does not readily unveil its true mathematical implications concerning the stability of the solutions or other aspects. Thus, from this point of view, the problem is described as improper or ill-posed. The most rigorous methods for solving the deconvolution equation are: regularization, spline function approximation and Fourier transform technique. The essential feature of regularization method is the replacement of a given improper problem with another, auxiliary, correctly posed problem. The second method consists in approximating both the experimental and instrumental signals by piecewise cubic spline. Most often when using this technique, the true sample function belongs to the same piecewise cubic spline class. The topic of this chapter is the application of Fourier transform in experimental data deconvolution for use in nanomaterial structures.

2. Mathematical background of signal deconvolution

As mentioned in the introduction, ill-posed problems from a mathematical point of view have many applications in physics and technologies [1]. In addition to the abovementioned examples, the examples below should be noted.

Solving the Cauchy problem for the Laplace equation, $\Delta U = 0$ has a direct application in biophysics as in [2]. The problem consists in determining the biopotential distribution within the body denoted by U, when the body surface potential values are known. The phenomenon is modelled by the Laplace equation, and the Cauchy conditions are $U \mid_S = f(S)$ and $\left. \frac{\partial U}{\partial n} \right|_S = 0$, where S represents the surface of the body.

The determination of radioactive substances in the body, as in [2], and protein crystallography structures also deal with ill-posed problems: see [3].

The same formalism is used in quantum mechanics to determine the particle scattering cross-section on different targets, as well as in plasma physics in the case of the electron distribution after speed is received from the dispersion curve analysis [2].

An intuitive way of grasping an ill-posed problem can be modelled by the movement of the vibrating string when many forces are acting perpendicularly on the string, as represented in Figure 1.

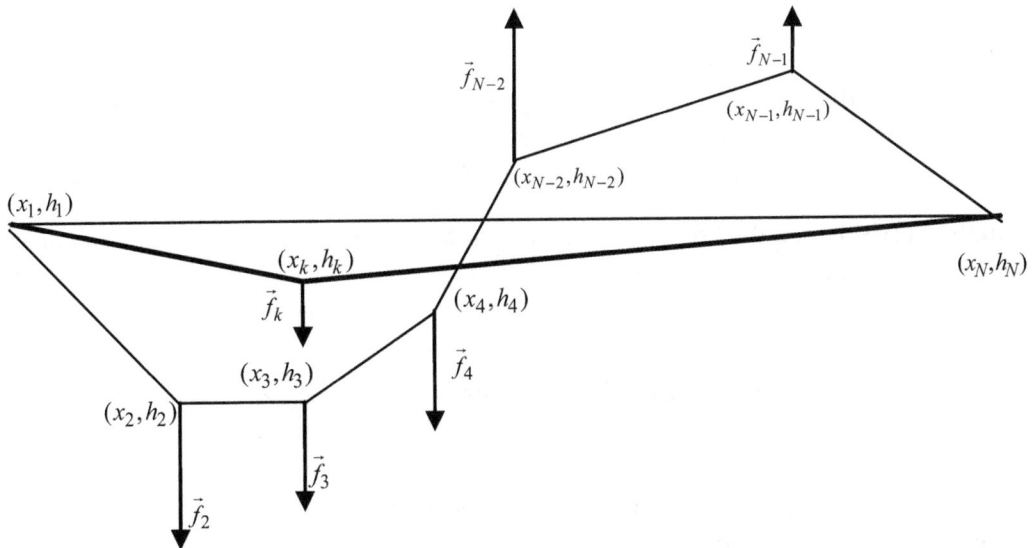

Figure 1. Physical model of the vibrating string

In terms of the mathematical equation, the phenomenon described above has a correspondent in physics spectroscopy used in the study of nanomaterials. In the first instance, it is considered that in the point of abscissa x_k, force $\vec{f}_k$ acts perpendicularly to the direction of the string. The string movement in the vertical plane at an arbitrary point s is given by the proportionality relationship,

$$h(s) = g(s, x_k) f(x_k)$$ (1)

where $g(s, x_k)$ characterizes the impact of force $\vec{f}(x_k)$ on the movement $h(s)$. Using the same considerations, for N forces $\vec{f}_1, \vec{f}_2, \cdots, \vec{f}_N$ that act independently in N points of abscissa $x_1, x_2, \ldots, x_N$ in the direction perpendicular to the string, the string movements will be obtained as $h_1, h_2, \ldots, h_N$. Therefore the movement associated with an arbitrary point of the string of abscissa s is described by the relation below

$$h(s) = g(s, x_1) f(x_1) + g(s, x_2) f(x_2) + \cdots + g(s, x_k) f(x_k) = \sum_{i=1}^{N} g(s, x_k) f(x_k)$$ (2)

When a force is distributed continuously along the entire string, the movement of the point s of the string will be given by

$$h(s) = \int_0^l g(s, x) f(x) dx$$ (3)

where l represents the length of the string.

The function f is the density of force, which means the force per unit length, and $f(x)dx$ represents the force that acts on the arc element dx. The function g is called the influence function because it shows the degree of influence of the distribution force f on displacement h.

The equation (3) is named the Fredholm integral equation of the first kind, and it is a particular case of the integral equation,

$$l(s) f(s) + \int_a^b g(s, x) f(x) dx = h(s), \quad c \le x \le d$$ (4)

where l, g and h are continuous known functions. If function l is null, then equation (4) represents an integral equation of the first kind. If function l has no zero on $[c, d]$ then the equation (4) is of the second kind, while if l has some zeros on $[c,d]$ then the equation (4) is of the third kind.

Although the aim of this chapter is signal deconvolution using the Fourier transform, it is important to mentionthe other two methods used to solve equation (4) when $l \equiv 0$, that is, the regularization method and the spline approach.

Hadamard stated that a problem is well posed if it has a unique solution, and the solution depends continuously on the data [4]. Any problem that is not a well-posed problem is an ill-posed one. The Fredholm equation of the first kind is ill posed because small changes in the data generate huge modification of the unknown function.

2.1. Regularization method

This method consists in the replacement of the ill-posed problem (4) with $l \equiv 0$ by a well-posed problem, and there are many scientific papers that develop different types of regularization method depending on kernel type and other specific needs. Below we describe the Tikhonov regularization method applied to the equation (4) with $l \equiv 0$. Let X and Y be Hilbert spaces and $\| \cdot \|$ be the norm on Hilbert space. If the kernel g is smooth, the operator $G : X \rightarrow Y$

$$(Gf)(s) = \int_a^b g(s,x) f(x) dx \qquad (5)$$

is linear. Then equation (4) with $l \equiv 0$ becomes

$$Gf = h \qquad (6)$$

The regularization method consists in the determination of the approximate solution of the equation (4) of the first kind as a minimization of the following functional

$$\Phi_a(f) = \| G - h \|^2 + \alpha \| Lf \|^2, \forall f \in X \qquad (7)$$

The value $\alpha > 0$ represents the regularization parameter and L is a linear operator defined below

$$Lf = a_0(f - \hat{f}) + a_1 f' + a_2 f'' \qquad (8)$$

where a_i has the value 0 or 1; and f' and f'' are the first and the second derivative of f. Function $\hat{f}$ represents a trial solution for equation (4) with $l \equiv 0$. The regularization order for the operator L is the same as the derivability order of f. The regularization parameter should be chosen carefully, because a good minimum for the functional (7) does not always lead to an adequate solution for equation (4) with $l \equiv 0$ as in [4]. The discrimination procedure of the equation (4) with $l \equiv 0$ and functional (7) depends on the specifics of each type of problem such as domain, type of kernel, etc., but this is not the subject of this chapter: see [4-6].

Some disadvantages of the regularization method, which is an iterative method, are the fact that it is very sensitive to the noise present in the experimental function and is time consuming.

If the kernel g of the equation (4) with $l=0$ has a delayed argument then the equation is called a convolution equation, and is widely applied in physical spectroscopy. The general form of a convolution equation is given by

$$h(s) = \int_{-\infty}^{\infty} g(s-x)f(x)dx \qquad (9)$$

After the change of variable $x=t-s$ it is found that (9) is equivalent by the equation

$$h(s) = \int_{-\infty}^{\infty} g(t)f(t-s)dt \qquad (10)$$

2.2. Spline technique

Spline functions for signal deconvolution technique help eliminate the drawbacks mentioned above [7]. The advantage of the method proposed in [7] lies in the fact that Beniaminy's method is a one-step method. In this case, the true sample function f is represented as a piecewise cubic spline function, and after the substitution of it into equation (10), the experimental function h becomes a piecewise cubic spline function with the same knots but different coefficients. The connection between the coefficients of functions h and f are given by the moments of instrumental function g. Thus, if function f has the form

$$f(t) = \sum_{k=1}^{n-1} s_k(t) \qquad (11)$$

where

$$s_k(t) = \begin{cases} 0 & \text{if} \quad t < \xi_k \\ a_k t^3 + b_k t^2 + c_k t + d_k & \text{if} \quad \xi_k \leq t \leq \xi_{k+1} \\ 0 & \text{if} \quad \xi_{k+1} < t \end{cases} \qquad (12)$$

with ξ_k, $k = \overline{1, n-1}$ are the knots and a_k, b_k, c_k and d_k are the coefficients of the spline function f. They are chosen such that the function together with its first two derivatives is continuous. Replacing (12) in (11), the experimental function has the form

$$h(s) = \sum_{k=1}^{n-1} \left(A_k t^3 + B_k t^2 + C_k t + D_k \right) \qquad (13)$$

where

$$
\begin{aligned}
A_k &= a_k M_0 \\
B_k &= b_k M_0 - 3 a_k M_1 \\
C_k &= 3 a_k M_2 - 2 b_k M_1 + c_k M_0 \\
D_k &= -a_k M_3 + b_k M_2 - c_k M_1 + d_k M_0
\end{aligned}
\tag{14}
$$

and $M_k = \int_{-\infty}^{\infty} t^k g(t)\,dt$ represents the moment of order k. Beniaminy considered that experimental function h given by (14) is a cubic spline function. In [8] it is shown that that function h is not a spline function due to the lack of the continuity in the first two derivatives of h. However, the algorithm from [7] gives good results, but the quality of the true sample function depends on how wide the instrumental function is. In order to obtain the true sample function f given by (12) and (13), we calculate spline coefficients of the experimental function and using these values and (14) obtain the coefficients a_k, b_k, c_k and d_k.

2.3. Solving the convolution equation using Fourier transform

Take the functions h, f, and g whose Fourier transform is given by the functions H, F and G. By applying the Fourier transform operator on both members of equation (10) we obtain

$$
\int_{-\infty}^{+\infty} h(s)\exp(-2\pi i v s)\,ds = \int_{-\infty}^{+\infty}\left[\int_{-\infty}^{+\infty} g(\tau)f(s-\tau)\,d\tau \exp(-2\pi i v s)\right]ds
\tag{15}
$$

By changing the order of integration, the Fourier transform of the signal h is expressed by the relation,

$$
H(v) = \int_{-\infty}^{+\infty} g(\tau)\left[\int_{-\infty}^{+\infty} f(s-\tau)\exp(-2\pi i v s)\,ds\right]d\tau
\tag{16}
$$

Using the substitution $\sigma = s - v$, the quantity between square brackets from the previous relation becomes

$$
\int_{-\infty}^{+\infty} f(\sigma)\exp\left[-2\pi i v(\sigma+\tau)\right]d\sigma = \exp(-2\pi i v\tau)\int_{-\infty}^{+\infty} f(\sigma)\exp(-2\pi i v\sigma)\,d\sigma = \exp(-2\pi i v\tau)F(\upsilon)
\tag{17}
$$

In this context the relation (16) becomes

$$
H(v) = \int_{-\infty}^{+\infty} g(\tau)\exp(-2\pi i v\tau)F(v)\,d\tau = F(v)G(v)
\tag{18}
$$

The relation (18) is known as the convolution theorem. If direct and inverse Fourier transform operators and convolution product are respectively denoted by TF, TF^{-1} and *, then the relation (18) is written symbolically as

$$TF(h) = TF(f)TF(g) = F\,G = TF(f * g)$$

and

$$TF^{-1}(F\,G) = h = f * g$$

In this way, the process of the inverse Fourier transform applied to function F determines f signal. In X-ray diffraction theory this is known as the Stokes method.

Experimental signals h, coded by (1), (3), (5) and (6) for a set of supported gold catalyst (Au/SiO$_2$), and instrumental contribution g measured on a gold foil, are presented in Figure 2.

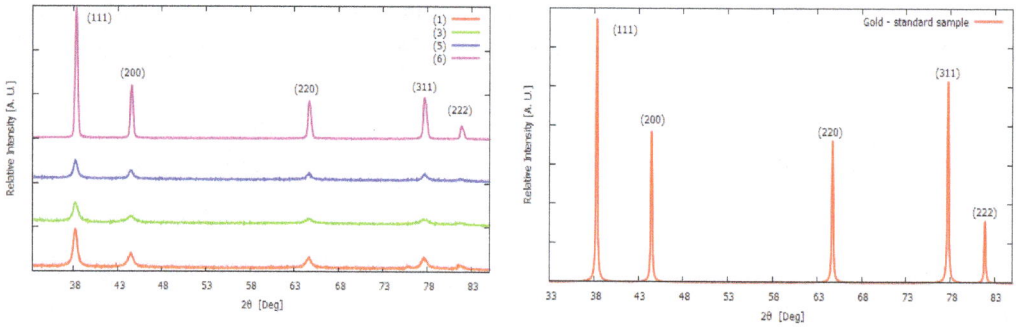

Figure 2. The experimental relative intensities h of the supported gold catalysts and instrumental function g

3. Why is the technique of deconvolution used in nanomaterials science?

In the scientific literature we can see many authors display serious confusion about the concept of deconvolution. Often, when they decompose the experimental signal h according to certain specific criteria, some say that it has achieved the deconvolution of the initial signal. This fact may be accepted only if the instrumental function g from equation (11) is described by the Dirac distribution. Only in this case is the true sample function f identical to the experimental signal h. Unfortunately, no instrumental function of any measuring device can be described by the Dirac distribution.

It is well known that the macroscopic physical properties of various materials depend directly on their density of states (DS). The DS is directly linked to crystallographic properties. For physical systems that belong to the long order class, moving the crystallographic lattice in the whole real space will reproduce the whole structure. The nanostructured materials, which belong to the short-range class, are obtained by moving the lattice in the three crystallographic directions at the limited distances, generating crystallites whose size is no greater than a few hundred angstroms. In this case, the DS is drastically modified in comparison with the previous class of materials. From a physical point of view the DS is closely related to the nanomaterials' dimensionality, so crystallite size gives direct information about new topological properties. It can emphasize that amorphous, disordered or weak crystalline materials can have new bonding and anti-bonding options. The systems consisting of nanoparticles whose dimensions do not exceed 50 Å have the majority of atoms practically situated on the surface for the most part. Additionally, the behaviour of crystallites whose size is between 50 Å and 300 Å is described on the basis of quantum mechanics to explain the advanced properties of the tunnelling effect. All these reasons lead to the search for an adequate method to determine reliable information such as effective particle size, microstrains of lattice, and particle distribution function. This information is obtained by Fourier deconvolution of the instrumental and experimental X-ray line profiles (XRLP) approximated by Gauss, Cauchy and Voigt distributions and generalized by Fermi function (GFF) as in [9]. The powder reflection broadening of the nanomaterials is normally caused by small size, crystallites and distortions within crystallites due to dislocation configurations. It is the most valuable and cheapest technique for the structural determination of crystalline nanomaterials.

Generally speaking, in X-ray diffraction on powder, the most accurate and reliable analysis of the signals is given by the convolution equation (10) where h, g and f are experimental data, instrumental contribution of setup experimental spectrum, and true sample function as a solution of equation (10), respectively.

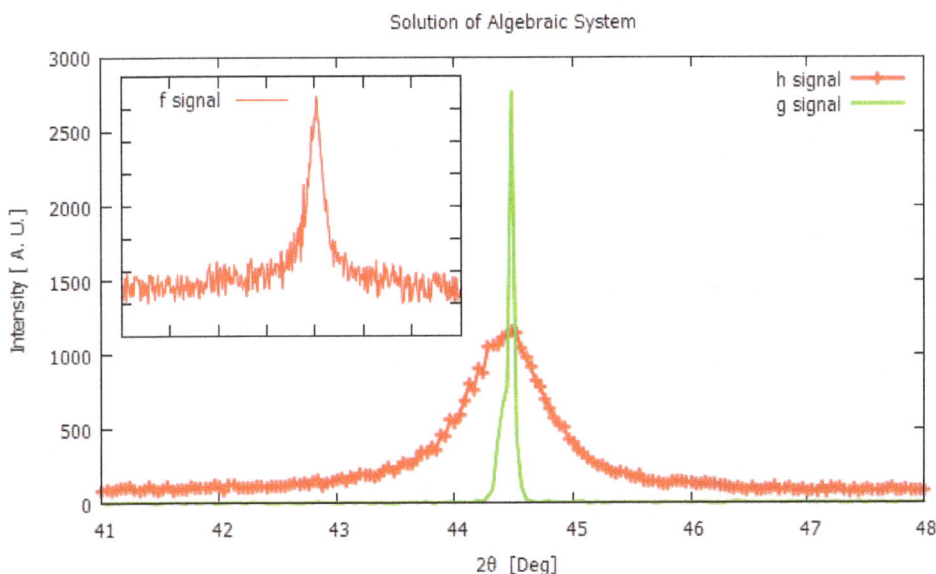

Figure 3. Numerical solution of the deconvolution equation (19) determined by an algebraic discretization

Let us consider the experimental signals of (111) X-ray line profile of supported nickel catalyst, and the instrumental function given by nickel foil obtained by a synchrotron radiation setup at 201 points with a constant step of 0.04^0 in 2θ variables, as shown in Figure 3.

The convolution equation (9) can be approximated in different ways, but the simplest approximation is given by following the algebraic system

$$h(x_i) = \sum_{j=-201}^{201} g(x_i - s_j) f(s_j) \Delta s_j \ldots i = 1,201 \tag{19}$$

where Δs_j is a constant step in 2θ variables. It turns out that the roots $f(s_j)$ of system (19) do not lead to a smooth signal, but yield a curve which makes for enhanced oscillations. Its behaviour is given by f signal in Figure 3. This result is given by a computer code written in Maple 11 language, a sequence of which is presented in Appendix 1. From a physical point of view, this type of solution is impracticable because the crystallite size in nanostructured systems is contained in the tails of XRLP. Therefore, the lobes of the XRLP must be sufficiently smooth. As shown in the inset of Figure 3, this condition is not met.. It would be possible to improve the quality of signal f trying to extend the definition interval for signal g. Thus we will approximate the unbounded integral on a bounded interval, but one that is sufficiently large.

This depends on the performances of the computer system and on the algorithm developed for solving inhomogeneous systems of linear equations with sizes of at least several thousand.

4. Distributions frequently used in physics and chemical signal deconvolution applied in nanomaterials science

It is known that, from a mathematical point of view, the XRLP are described by the symmetric or asymmetric distributions. As in [10,11] a large variety of functions for analysis of XRLP, such as Voigt (V), pseudo-Voigt (pV) and Pearson VII (P7), are proposed.

4.1. Gauss distribution

Many results such as the propagation of uncertainties and the least square method can be derived analytically in explicit form when the relevant variables are normally distributed. Gauss distribution is defined by mathematical relation

$$I_G = \frac{I_{0G}}{\sqrt{\pi}\gamma_G} \exp\left[-\left(\frac{x-a}{\gamma_G}\right)^2\right] \tag{20}$$

where I_{0G}, a and γ_G are the profile area, gravitational centre measured in 2θ variable, and broadening of the XRLP, respectively. The nth moment, $n=0,1$ is given by relations

$$\mu_{0G} = \int_{-\infty}^{+\infty} I_G(x)dx = I_{0G}, \qquad \mu_{1G} = a$$

The integral width δ_G and full width at half maximum $FWHM_G$ are given by relations

$$\delta_G = \sqrt{\pi}\gamma_G \quad \text{and} \quad FWHM_G = 2\sqrt{\ln 2}\gamma_G$$

If both signals h and g are described by Gaussian distributions and take into account the relationship (18), the full width and FWHM of the true sample function are expressed by the relations

$$\gamma_{G,f} = \sqrt{\gamma_{G,h}^2 - \gamma_{G,g}^2} \quad FWHM_{G,f} = 2\sqrt{\ln 2}\gamma_{G,f} \tag{21}$$

4.2. Cauchy distribution

The Cauchy distribution, also called the Lorentzian distribution, is a continuous distribution that describes population distribution of electron levels with multiple applications in physical spectroscopy. Its analytical expression is given by relation

$$I_C = \frac{I_{0C}}{\pi} \frac{\gamma_C}{\gamma_C^2 + (x-a)^2} \tag{22}$$

where I_{0C}, a and γ_C are profile surface, gravitational centre and broadening of the XRLP, respectively. The nth moment $n=0,1$ is given by relations

$$\mu_{0C} = \int_{-\infty}^{+\infty} I_C(x)dx = I_{0C} \quad \text{and} \quad \mu_{1C} = a$$

The integral widths δ_G and full width at half maximum $FWHM_C$ are given by relations

$$\delta_C = \pi\gamma_C \quad \text{and} \quad FWHM_C = 2\gamma_C$$

The deconvolution of two signals h and g determined by Cauchy distributions is also a Cauchy distribution whose full width $\delta_{C,f}$ and $FWHM_{C,f}$ are given by relations

$$\delta_{C,f} = \delta_{C,h} - \delta_{C,g} \quad \text{and} \quad FWHM_{C,f} = 2\delta_{C,f} \tag{23}$$

4.3. Generalized Fermi function

Although extensive research over the past few decades has made progress in XRLP global approximations, their complete analytical properties have not been reported in the literature. Unfortunately, most of them have complicated forms, and they are not easy to handle mathematically. Recently, as in [9,11], a simple function with a minimal number of parameters named the generalized Fermi function (GFF), suitable for minimization and with remarkable analytical properties, was presented from a purely phenomenological point of view. It is given by the relationship,

$$h(s) = \frac{A}{e^{-a(s-c)} + e^{b(s-c)}} \tag{24}$$

where A, a, b, c are unknown parameters. The values A, c describe the amplitude and the position of the peak, and a, b control its shape. If $b=0$, the h function reproduces the Fermi-Dirac electronic energy distribution. The GFF has remarkable mathematical properties, with direct use in determining the moments, the integral width, and the Fourier transform of the XRLP, as well as the true sample function. Here we give its properties without proofs.

i.   By setting

$$s' = s - c \quad \rho = (a+b)/2 \quad q = (a-b)/2$$

we obtain

$$h(s') = \frac{A}{2}\left(\frac{\cosh qs' + \sinh qs'}{\cosh \rho s'} \right) \tag{25}$$

ii.   the limit of h function for infinite arguments is finite, so $\lim h(s')=0$ when $s' \to \pm\infty$;

iii.   the zero, first and second order moments (μ_0, μ_1, μ_2) of the h function are given by the relations

$$\mu_0 = \frac{\pi A}{2\rho \cos\frac{\pi q}{2\rho}}, \quad \mu_1 = \frac{\pi}{2\rho}\tan\frac{\pi q}{2\rho}, \quad \mu_2 = \left(\frac{\pi}{2\rho}\right)^2\left(\frac{1}{\cos^2\frac{\pi q}{2\rho}} + \tan^2\frac{\pi q}{2\rho} \right)$$

iv.   the integral width $\delta_h(a, b)$ of the h function has the following form

$$\delta_h(a,b) = \frac{\pi}{\left(a^a b^b\right)^{1/(a+b)} \cos\left(\frac{\pi}{2}\frac{a-b}{a+b}\right)} \tag{26}$$

v.   the Fourier transform of the h function is given by the relationship

$$H(L) = \frac{A}{2}\int_{-\infty}^{+\infty}\frac{\cosh qs' + \sinh qs'}{\cosh \rho s'}e^{-2\pi s'L}\,ds' = \frac{\pi A}{2\rho\left|\cos\left(\frac{\pi q}{2\rho}+i\frac{\pi^2 L}{\rho}\right)\right|^2}\cos\left(\frac{\pi q}{2\rho}-i\frac{\pi^2 L}{\rho}\right)$$

(27)

vi.   if we consider the functions f and g defined by equation (25), by their deconvolution we can compute the $|F(L)|$ function, which is used in Warren and Averbach's analysis in [12]. Therefore, the magnitude of $F(L)$ function has the following form:

$$|F(L)| = \frac{A_h\rho_g}{A_g\rho_h}\sqrt{\frac{\cos^2\alpha + \sinh^2\beta L}{\cos^2\gamma + \sinh^2\delta L}},$$

(28)

where the arguments of trigonometric and hyperbolic functions are expressed by

$$\alpha = \frac{\pi q_g}{2\rho_g},\quad \beta = \frac{\pi^2}{\rho_g},\quad \gamma = \frac{\pi q_h}{2\rho_h},\quad \delta = \frac{\pi^2}{\rho_h}$$

The subscripts g and h refer to the instrumental and experimental XRLP. Taking into account the convolution theorem, the true sample function f is given by the relationship

$$f(s) = \frac{A_h\rho_g}{A_g\rho_h}\int_{-\infty}^{+\infty}\frac{\cos\left(\frac{\pi q_g}{2\rho_g}+i\frac{\pi^2 L}{\rho_g}\right)}{\cos\left(\frac{\pi q_h}{2\rho_h}+i\frac{\pi^2 L}{\rho_h}\right)}\exp(2\pi iLs)\,ds$$

The last integral cannot be accurately resolved. In order to do so we have to consider some arguments. The Fourier transform of f is the F function, given by the relations

$$F(L) = |F(L)|\exp(i\theta(L)),\quad \theta(L) = \arctan\frac{\Im(F(L))}{\Re(F(L))}$$

where θ means the angle function, and $\Re(F)$ and $\Im(F)$ are the real and imaginary parts of the complex function F, respectively. The arguments α, β, γ and δ from equation (28) depend only on the asymmetry parameters a and b of the g and f functions. If the XRLP asymmetry is not very large (i.e., a and b parameters are close enough as values) the $\cos^2\alpha\approx1$, $\cos^2\gamma\approx1$ approximations are reliable. Therefore, we obtain $\Im(F)<<\Re(F)$, $\theta(L)\approx0$ and the magnitude of the Fourier transform for the true XRLP sample can be expressed as

$$|F(L)| = \frac{A_h \rho_g}{A_g \rho_h} \frac{\cosh \dfrac{\pi^2 L}{\rho_g}}{\cosh \dfrac{\pi^2 L}{\rho_h}} \tag{29}$$

(vii) if we consider the previous approximation, the true XRLP sample is given by an inverse Fourier transform of the F function, and consequently we have

$$f(s') = \frac{2 A_h \rho_g}{\pi A_g} \frac{\cos \frac{\pi \rho_h}{2 \rho_g} \cosh \rho_h s'}{\cosh 2 \rho_h s' + \cos \frac{\pi \rho_h}{\rho_g}} \tag{30}$$

(viii) the integral width of the true XRLP sample can be expressed by the δ_f function

$$\delta_f(\rho_h, \rho_g) = \frac{\pi}{2 \rho_h \cos \frac{\pi \rho_h}{2 \rho_g}} \left(\cos \frac{\pi \rho_h}{\rho_g} + 1 \right) \tag{31}$$

4.4. Voigt distribution applied in X-ray line profile analysis

Before briefly describing the mathematical properties of the Voigt distribution, let us examine the physical concept underlying the approximation of the XRLP by Voigt distribution and the convolution process.

During decades of research, Warren and Averbach [12] introduced the X-ray diffraction concept for the mosaic structure model, in which the atoms are arranged in blocks, each block itself being an ideal crystal, but with adjacent blocks that do not accurately fit together. They considered that the XRLP h represents the convolution between the true sample f and the instrumental function g, produced by a well-annealed sample. The effective crystallite size D_{eff} and lattice disorder parameter $<\varepsilon_{hkl}>$ were analysed as a set of independent events in a likelihood concept. Based on Fourier convolution produced between f and g signals and the mosaic structural model, the analytical form of the Fourier transform for the true sample function was obtained. The normalized F was described as the product of two factors, $F^{(s)}(L)$ and $F^{(\varepsilon)}(L)$, where variable L represents the distance perpendicular to the (hkl) reflection planes. The factor $F^{(s)}(L)$ describes the contribution of crystallite size and stocking fault probability, while the factor $F^{(\varepsilon)}(L)$ gives information about the microstrain of the lattice. The general form of the Fourier transform of the true sample for cubic lattices was given by relationships

$$F^{(s)}(L) = e^{-\frac{|L|}{D_{eff}(hkl)}}, \qquad F^{(\varepsilon)}(L) = e^{-\frac{2\pi^2 \langle \varepsilon_L^2 \rangle_{hkl} h_0^2 L^2}{a^2}}, \tag{32}$$

where $h_0^2 = h^2 + k^2 + l^2$. The general form of the true sample function f is given by inverse Fourier transform of $F(L)$

$$f(s) = \int_{-\infty}^{\infty} e^{-\beta L^2 - \gamma |L|} e^{2\pi i s L} dL =$$

$$= \sqrt{\frac{\pi}{\beta}} \exp\left[\frac{\gamma^2 - (2\pi s)^2}{4\beta}\right]\left\{\Re\left(erfc\left(\frac{\gamma - 2\pi i s}{2\sqrt{\beta}}\right)\right)\cos\frac{\pi\gamma s}{\beta} - \Im\left(erfc\left(\frac{\gamma + 2\pi i s}{2\sqrt{\beta}}\right)\right)\sin\frac{\pi\gamma s}{\beta}\right\}$$

(33)

where $s = 2\left(\dfrac{\sin\theta}{\lambda} - \dfrac{\sin\theta_0}{\lambda}\right)$, $erf(x) = \dfrac{2}{\sqrt{\pi}}\int_0^x e^{-t^2}dt$, $erfc(x) = 1 - erf(x)$ is the complementary error

function [13] and $\beta = \dfrac{2\pi^2 \langle \varepsilon_L^2 \rangle_{hkl} h_0^2}{a^2}$, $\gamma = \dfrac{1}{D_{eff}(hkl)}$. The last relation from the mathematical

point of view represents a Voigt distribution. If we take into account the properties of the Gauss and Cauchy distributions, the Voigt distribution can be generalized by relation

$$V(x, \gamma_G, \gamma_C) = \int_{-\infty}^{+\infty} I_G(x', \gamma_G, x_{0G}) I_C(x - x', \gamma_C, x_{0C})dx'$$

(34)

Based on relation (18), its Fourier transform is given by $FT[V] = \underset{F_G}{FT[I_G]} \cdot \underset{F_C}{FT[I_C]}$ where

$$F_G(L) = e^{-2\pi i x_{0G}L_e - \pi^2\gamma_G^2 L^2} \quad \text{and} \quad F_C(L) = e^{-2\pi i x_{0C}L - 2\pi\gamma_C|L|}$$

The analytical expression of the Voigt distribution is

$$V(x, \gamma_G, \gamma_C) = FT^{-1}\left[e^{-2\pi i(x_{0C} + x_{0G})L} e^{-\pi^2\gamma_G^2 L^2 - 2\pi\gamma_C|L|}\right]$$

(35)

Explicit forms of experimental signal and true sample function normalized at I_{0V} are given by relations [13]

$$V(x, \gamma_G, \gamma_C) = \frac{I_{0V}}{\sqrt{\pi}\gamma_G}\exp\left[\frac{\gamma_C^2 - (x - x_{0C} - x_{0G})^2}{\gamma_G^2}\right]\cdot$$

$$\cdot\left\{\Re\left[erfc\left(\frac{\gamma_C - i(x - x_{0C} - x_{0G})}{\gamma_G}\right)\right]\cos\frac{2\gamma_C(x - x_{0C} - x_{0G})}{\gamma_G^2} -\right.$$

$$\left. - \Im\left[erfc\left(\frac{\gamma_C + i(x - x_{0C} - x_{0G})}{\gamma_G}\right)\right]\sin\frac{2\gamma_C(x - x_{0C} - x_{0G})}{\gamma_G^2}\right\}$$

(36)

and

$$
\begin{aligned}
V_f(x, \gamma_{G,f}, \gamma_{C,f}) = {} & \frac{1}{\sqrt{\pi}\gamma_{C,f}} \exp\left[\frac{\gamma_{C,f}^2 - (x - x_{0C} - x_{0G})^2}{\gamma_{G,f}^2}\right] \cdot \\
& \cdot \left\{ \Re\left[erfc\left(\frac{\gamma_{C,f} - i(x - x_{0C} - x_{0G})}{\gamma_{G,f}}\right)\right] \cos\frac{2\gamma_{C,f}(x - x_{0C} - x_{0G})}{\gamma_{G,f}^2} - \right. \\
& \left. - \Im\left[erfc\left(\frac{\gamma_{C,f} + i(x - x_{0C} - x_{0G})}{\gamma_{G,f}}\right)\right] \sin\frac{2\gamma_C(x - x_{0C} - x_{0G})}{\gamma_{G,f}^2} \right\}
\end{aligned}
\tag{37}
$$

Maximum value of true sample function is

$$
V_{max} = \frac{1}{\sqrt{\pi}\gamma_G} \exp\left(\frac{\gamma_C^2}{\gamma_G^2}\right) erfc\left(\frac{\gamma_C}{\gamma_G}\right)
$$

Voigt function is a probability density function and it displays the distribution of target values,

$$
\int_{-\infty}^{\infty} V(x, \gamma_G, \gamma_C)dx = \frac{1}{\pi\sqrt{\pi}} \frac{\gamma_C}{\gamma_G} \int_{-\infty}^{\infty}\left[\int_{-\infty}^{\infty} \frac{e^{-\left(\frac{x'}{\gamma_G}\right)^2}}{\gamma_C^2 + (x - x')^2} dx'\right] dx = 1,
$$

$$
\delta_V = \frac{\sqrt{\pi}\gamma_G}{\exp\left(\frac{\gamma_C}{\gamma_G}\right)^2 erfc\left(\frac{\gamma_C}{\gamma_G}\right)}
\tag{38}
$$

and the convolution of two Voigt functions is also a Voigt function.

The integral width of a true sample function has the two components given by the Gauss and Cauchy contributions

$$
\delta_{G,f}^2 = \frac{1}{\pi}\left(\delta_{G,h}^2 - \delta_{G,g}^2\right), \quad \delta_{C,f} = \frac{1}{\pi}\left(\delta_{C,h} - \delta_{C,g}\right)
\tag{39}
$$

Balzar and Popa are among the leading scientists in the field of Fourier analysis of X-ray diffraction profiles, and they suggested that each Gauss and Cauchy component contains information about the average crystallite size (δ_S) and distortion of the lattice (δ_D) as in [14]. From the algebraic point of view, they proposed the following relationship

$$
\delta_{G,f}^2 = \delta_{SG,f}^2 + \delta_{DG,f}^2, \quad \delta_{C,f} = \delta_{SC,f} + \delta_{DC,f}
\tag{40}
$$

Based on the new concept introduced by them, the two components of the Fourier transform are given by the relations

$$F^S(L) = e^{-\pi L^2 \delta_{SG,f}^2 - 2|L|\delta_{SC,f}}, \quad F^D(L) = e^{-\pi L^2 \delta_{DG,f}^2 - 2|L|\delta_{DC,f}} \tag{41}$$

The particle size distribution function, $P(L)$ is determined from the second derivative of strain-corrected Fourier transform of the true sample function. The volume-weighted column-length P_V and surface-weighted column-length P_S distributions are given by the following [14]:

$$P_V(L) = L \frac{d^2 F^S(L)}{dL^2} = 2L \left[2\left(\pi L \delta_{SG}^2 + \delta_{SC}\right)^2 - \pi \delta_{SG}^2 \right] F^S(L) \tag{42}$$

$$P_S(L) = P_V(L) = \frac{d^2 F^S(L)}{dL^2} = 2 \left[2\left(\pi L \delta_{SG}^2 + \delta_{SC}\right)^2 - \pi \delta_{SG}^2 \right] F^S(L) \tag{43}$$

5. Experimental section, data analysis and results

A series of four supported gold catalysts were studied by X-ray diffraction (XRD) in order to determine the average particle size of the gold, the microstrain of the lattice as well as the size and microstrain distribution functions by XRLP deconvolution using Fourier transform technique. The gold catalyst samples with up to 5 wt% gold content were prepared by impregnation of the SiO_2 support with aqueous solution of $HAuCl_4 \times 3H_2O$ and homogeneous deposition-precipitation using urea as the precipitating agent method, respectively. The X-ray diffraction data of the supported gold catalysts displayed in Figure 3 were collected using a Rigaku horizontal powder diffractometer with rotated anode in Bragg-Brentano geometry with Ni-filtered Cu $K\alpha$ radiation, $\lambda = 1.54178$ Å, at room temperature. The typical experimental conditions were: 60 sec for each step, initial angle $2\theta = 32^0$, and a step of 0.02^0, and each profile was measured at 2700 points. The XRD method is based on the deconvolution of the experimental XRLP (111) and (222) using Fourier transform procedure by fitting the XRLP with the Gauss, Cauchy, GFF and Voigt distributions. The Fourier analysis of XRLP validity depends strongly on the magnitude and nature of the errors propagated in the data analysis. The scientific literature treated three systematic errors: uncorrected constant background, truncation, and effect of sampling for the observed profile at a finite number of points that appear in discrete Fourier analysis. In order to minimize propagation of these systematic errors, a global approximation of the XRLP is adopted instead of the discrete calculus. The reason for this choice was the simplicity and mathematical elegance of the analytical Fourier transform magnitude and the integral width of the true XRLP given by equations (20)-(24), (31), (34) and (38), as in [15]. The robustness of these approximations for the XRLP arises from the possibility of using the analytical forms of the Fourier transform instead of a numerical fast Fourier transform (FFT). It is well known that the validity of the numerical FFT depends drastically on the filtering technique what was adopted in [16]. In this way, the validity of the nanostructural parameters is closely related to the accuracy of the Fourier transform magnitude of the true XRLP.

Experimental relative intensities (111) with respect to 2θ values for (1) system are shown in Figure 4. The next steps consist in background correction of XRLP by polynomial procedures, finding the best parameters for the distributions adopted using the method of least squares or nonlinear fit, and then deconvoluting them using instrumental function. The main steps in the data analysis of the investigated systems are shown in Figure 4.

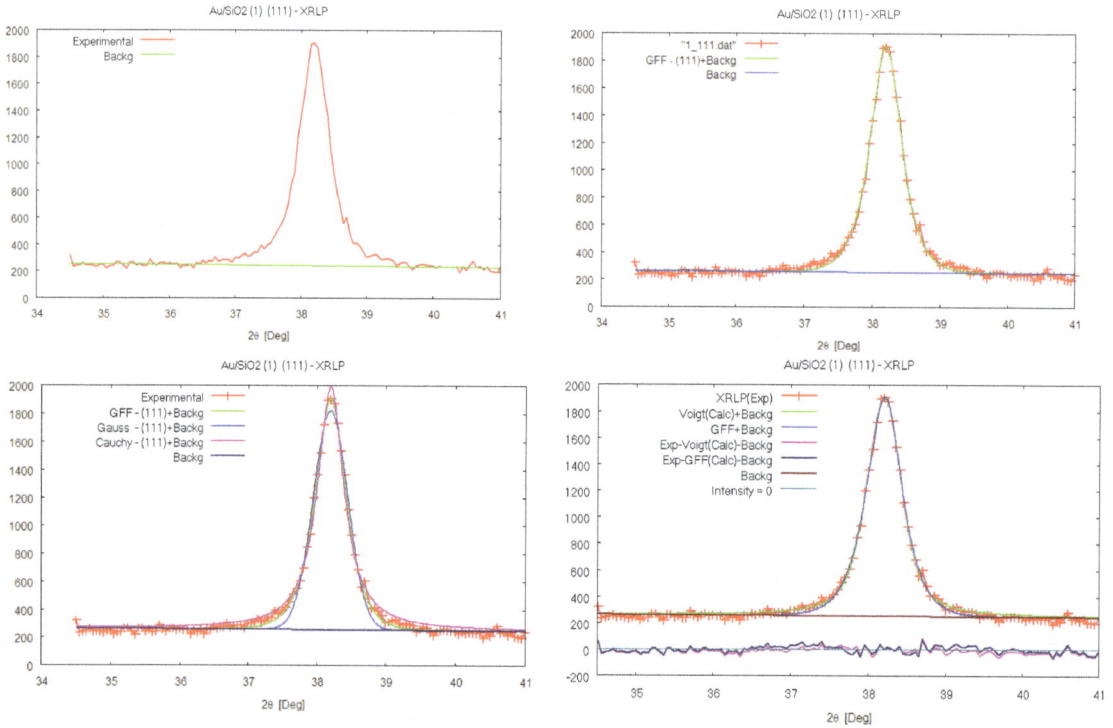

Figure 4. Various stages of processing for X-ray line profile (111) of the sample (1)

Experimental relative intensities (222) with respect to 2θ values for (6) system are shown in Figure 5.

The Fourier transforms normalized for the true sample function of the investigated samples (1) and (6) were calculated by three distinct methods, based on relations (28), (32) and (41), and are displayed in Figure 6.

The microstrain and particle size distribution functions determined by Fourier deconvolution of a single XRLP were calculated using equation (32), and are plotted in Figure 7.

The credibility of the parameters describing the investigated nanostructure systems depends primarily on the process of approximation of XRLP. This criterion is expressed by the root mean squares of residuals (*rmsr*) of data analysis and is given by relation

$$rmsr = 100\sqrt{\frac{\sum_{i=1}^{N_{points}} \frac{\left(y_i^{calc} - y_i^{exp}\right)^2}{\sigma_i^2}}{N_{points} - N_{param}}}$$

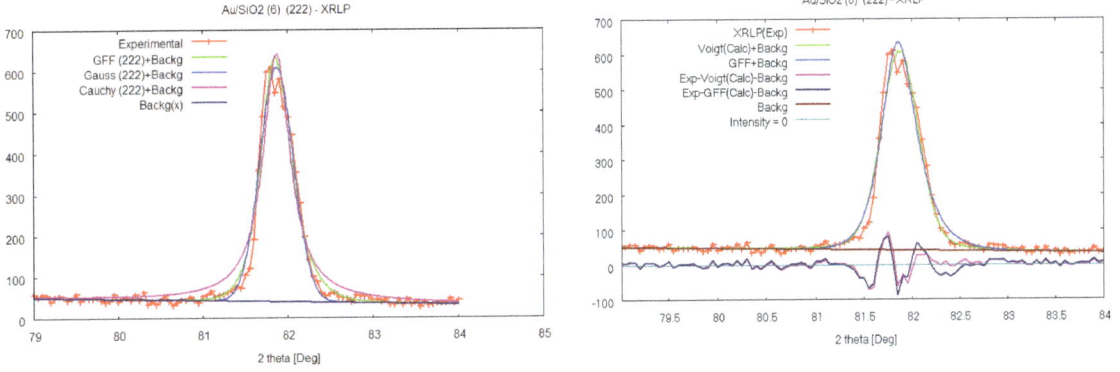

Figure 5. Various stages of processing for X-ray line profile (222) of the sample (6)

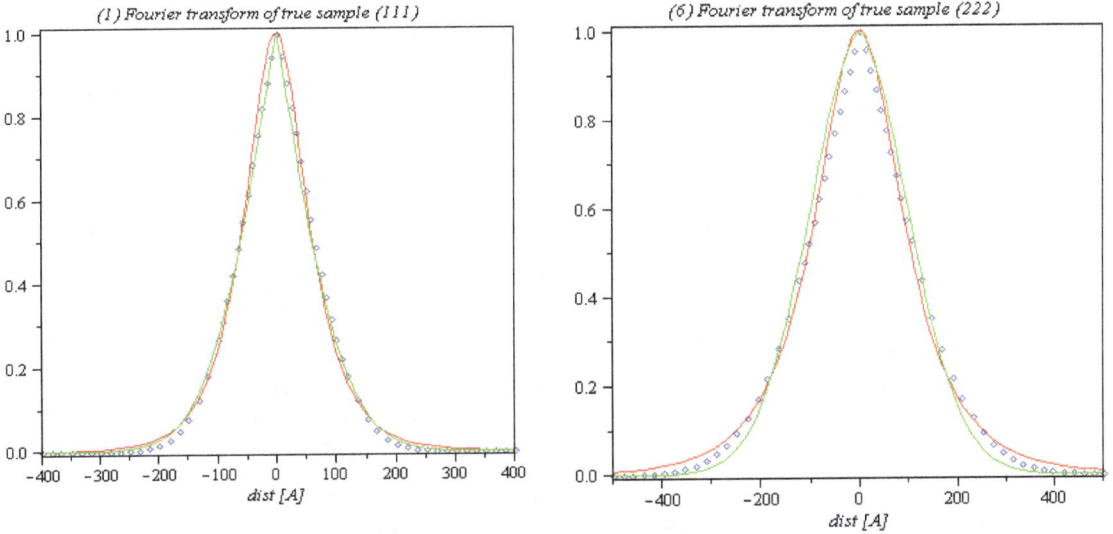

Figure 6. Fourier transform of true sample function of XRLP (111) and (222) for systems (1) and (6): blue - general relation, red - GFF, green - Voigt distribution

The *rmsr* values for all distributions used in XRLP approximation process are given in Table 1. The *rmsr* values are closely related to the spectral noise of experimental data. Here it is shown that a model based on GFF and Voigt distribution may be more realistic and accurate.

The integral widths and FWHM of the true sample functions calculated for all distributions were determined using the relations (21), (23), (31) and (38). Their values are presented in Table 2.

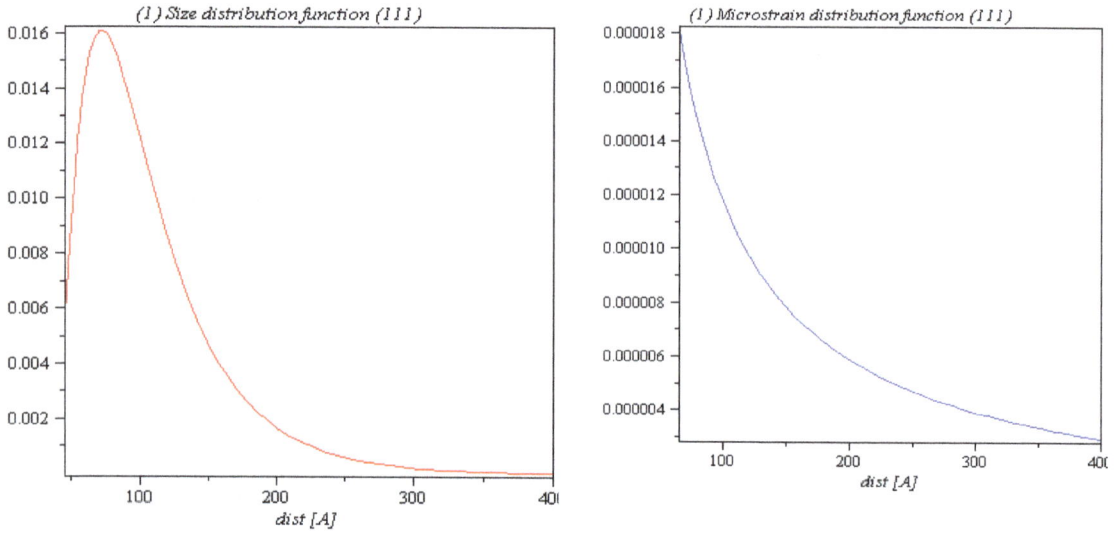

Figure 7. Size and microstrain distribution functions of (111) XRLP for system (1)

		Distribution			
Sample	hkl	GFF	Gauss	Cauchy	Voigt
(1)	111	26.2658	47.6477	40.6091	26.9398
	222	14.8795	15.2648	16.0429	15.1880
(3)	111	21.5743	31.4428	28.0142	22.1633
	222	12.5725	12.4615	12.6212	12.4534
(5)	111	18.1273	26.6941	21.8148	17.4276
	222	13.2033	13.3122	13.2599	13.2771
(6)	111	28.1267	31.8679	36.0040	35.5948
	222	18.8487	20.5945	33.2718	19.9656

Table 1. Values for *rmsr* for investigated samples

Because the experimental XRLP was measured for both (111) and (222), the surface-weighted column-length P_S and volume-weighted column-length P_V distribution functions were determined using relations (42,43) implemented in BREADTH software [17]. Additionally, it has found that the Gumbel distribution is the most adequate function for the global approximation of both probabilities' curves, and the results are shown in Figure 8.

The global structural parameters obtained for the investigated samples are summarized in Table 3 and Table 4.

Sample	hkl	$2\theta^0$ [Deg]	Distributions							
			GFF		Gauss		Cauchy		Voigt	
			δ [Deg]	a* [Deg]	δ [Deg]	a* [Deg]	δ [Deg]	a* [Deg]	δ [Deg]	a* [Deg]
(1)	111	38.291	0.644	0.540	0.638	0.599	0.738	0.469	0.713	0.537
	222	81.780	0.862	0.715	0.854	0.803	1.025	0.653	0.927	0.765
(3)	111	38.220	0.802	0.672	0.798	0.750	0.919	0.585	0.891	0.668
	222	81.763	1.028	0.860	1.011	0.950	1.181	0.752	1.095	0.912
(5)	111	38.211	0.631	0.529	0.621	0.583	0.723	0.460	0.701	0.519
	222	81.801	1.052	0.873	1.067	1.003	1.225	0.780	1.185	0.890
(6)	111	38.292	0.363	0.304	0.363	0.341	0.426	0.271	0.374	0.335
	222	81.872	0.506	0.423	0.498	0.468	0.613	0.390	0.528	0.461

a* represents FWHM

Table 2. Values for integral width and full width at half maximum for investigated samples

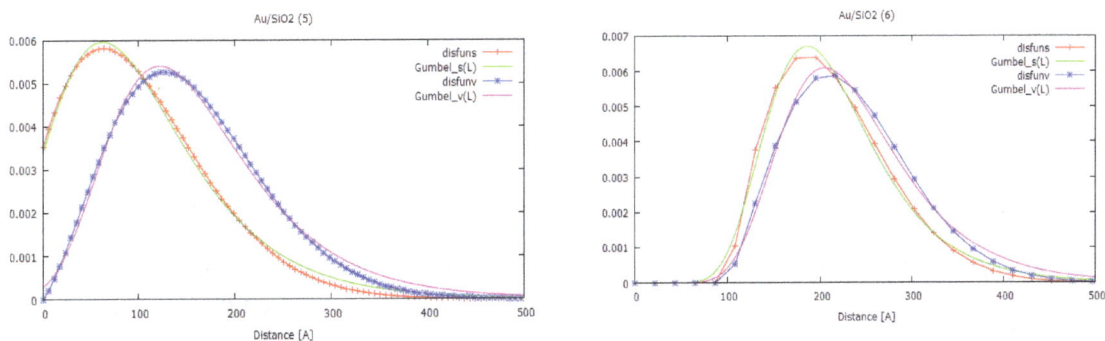

Figure 8. Surface-weighted column-length distribution function, P_S, and volume-weighted column-length distribution function, P_V, for (5) and (6) systems

Sample	GFF approximation				Single Voigt approximation			
	D_{111}^{Sch}	D_{111}	D_{222}^{Sch}	D_{222}	D_{111}^{Sch}	D_{111}	D_{222}^{Sch}	D_{222}
(1)	180	168	180	139	145	154	118	145
(3)	143	135	139	130	116	118	98	139
(5)	183	171	145	112	153	151	108	95
(6)	398	309	313	263	249	384	199	303

Table 3. Values for crystallite size determined by Scherrer method, and effective crystallite size using single XRLP approximations

Sample	hkl	$\langle D \rangle_S \pm \Delta D_S$ [Å]	$\langle D \rangle_V \pm \Delta D_V$ [Å]	$\langle \varepsilon^2 (D_V / 2) \rangle^{1/2} \pm \Delta \langle \varepsilon^2 (D_V / 2) \rangle^{1/2}$
(1)	[111/222]	90±12	133±15.	0.211E-02 ± 0.152E-03
(3)	[111/222]	69±13	104± 17	0.262E-02 ± 0.288E-03
(5)	[111/222]	106±30	153±41	0.280E-02 ± 0.216E-03
(6)	[111/222]	214±72	233±79	0.105E-02 ± 0.522E-03

Table 4. Values for the average crystallite size and microstrain using double Voigt approaches

Hydrogen chemisorption, transmission electron microscopy (TEM), magnetization, electronic paramagnetic resonance (EPR) and other methods could also be used to determine the average diameter of particles by taking into account a prior spherical form for the grains. By XRD method we can obtain the crystallite sizes that have different values for different crystallographic planes. There is a large difference between the particle size and the crystallite size due to the different physical meaning of the two concepts. It is possible that the particles of the supported gold catalysts are made up of many gold crystallites.

The size of the crystallites determined by equations (32) and (41), corresponding to (111) and (222) planes, have different values. The crystallite sizes D_{111}^{Sch} and D_{222}^{Sch} are determined by the Scherrer method [18] without taking into account the microstrain of the lattice. The values D_{111} and D_{222} were determined by Fourier deconvolution method for single XRLP, while the averages of D_V and D_S were calculated by a double Voigt approach. The difference between the crystallites' size can be explained by the fact that the analytical models are different due to the different approaches. This means that the geometry of the crystallites is not spherical [18]. The microstrain parameter of the lattice can also be correlated with the effective crystallite size in the following way: the value of the effective crystallite size increases when the microstrain value decreases.

The main procedures of the SIZE.mws software dedicated to Fourier analysis of the XRLP by GFF and Voigt distributions written in Maple 11 language are presented in Appendix 2.

6. Conclusions

In the present chapter, it is shown that XRD analysis provides more information for understanding the physical properties of nanomaterial structure. Powder X-ray diffraction is the cheapest and most reliable method compared with hydrogen chemisorptions, TEM techniques, magnetic measurements, EPR, etc. The main conclusions that can be drawn from these studies are:

1. For XRLP analysis, a global approximation should be applied rather than a numerical Fourier analysis. The former analysis is better than a numerical calculation because it can minimize the systematic errors that could appear in the traditional Fourier analysis.

2. Our numerical results show that by using the GFF and the Voigt distribution we successfully obtained reliable global nanostructural parameters;

3. Cauchy and Gauss distributions used for XRLP approximation give roughly structural information;

4. Powder X-ray diffraction gives the most detailed nanostructural results, such as: average crystallite size, microstrain, and distribution functions of crystallite size and microstrain;

5. Surface-weighted domain size depends only on Cauchy integral breadth, while volume-weighted domain size depends on Cauchy and Gauss integral breadths;

6. To obtain valid structural results, it is important to have: a good S/N ratio of the experimental spectra, a good deconvolution technique for the experimental and instrumental spectra, and an adequate computer package and programs for data analysis.

Appendix 1

Input data h.txt and g.txt files

```
`k`:=1;
line_h:= readline(`h.txt`):
line_g:= readline(`g.txt`):
while line <> 0 do
temp_h:= sscanf(line_h,`%8f%8f`):temp_g:=sscanf(line_g,`%8f %8f`):
printf(`%10.5f %10.5f`,temp_h[1],temp_h[2]): lprint():
printf(`%10.5f %10.5f`,temp_g[1],temp_g[2]): lprint():
twotheta_h[`k`]:=temp_h[1];
intensity_h[`k`]:=temp_h[2];
twotheta_g[`k`]:=temp_g[1];
intensity_g[`k`]:=temp_g[2];
line_h:= readline(`h.txt`):
line_g:= readline(`g.txt`):
`k`:=`k`+1;
end do;
`k`:=`k`-1;
p_h:=plot([twotheta_h[`ih`],intensity_h[`ih`],`ih`=1..k],col-
or=red,style=LINE,thickness=2,axes=boxed,gridlines,
labels=["2theta",""]):
p_g:=plot([twotheta_g[`ig`],intensity_g[`ig`],`ig`=1..k],col-
or=blue,style=LINE,thickness=2,axes=boxed,gridlines,
labels=["2theta",""]):
display({p_h,p_g});
deltatwotheta:=twotheta_h[2]-twotheta_h[1]:
```

h vector determination

```
twok:=2*k;
for `i` from 1 to k
do
h[`i`]:=intensity_h[`i`]:
end do:
```

```
h[`k`+1]:=0:
`j`:=1:
for `i` from `k`+2 to twok+1
do
h[`i`]:=intensity_h[`j`]:
`j`:=`j`+1:
end do:
print(h);
```

g array determination

```
`j`:=1:
for `i` from -twok to -k
do
g[`i`]:=intensity_g[`j`]:
`j`:=`j`+1;
end do:
`j`:=1:
for `i` from -k to -1
do
g[`i`]:=intensity_g[`j`]:
`j`:=`j`+1:
end do:
g[0]:=0.:
for `i` from 1 to k
do
g[`i`]:=intensity_g[`i`]:
end do:
`j`:=1:
for `i` from k+1 to twok
do
g[`i`]:=intensity_g[`j`]:
`j`:=`j`+1:
end do:
print(g):
```

a matrix determination

```
for `i` from 1 to twok+1
do
for `j` from 1 to twok+1
do
a[`i`,`j`]:=0.:
end do:
end do:
`i1`:=0:
for `i` from -k to k
do
`j1`:=0:
`i1`:=`i1`+1:
for `j` from -k to k
do
```

```
`j1`:=`j1`+1:
a[`i1`,`j1`]:=g[`i`-`j`]*deltatwotheta;
end do:
end do:
print(a);
```

solving integral deconvolution equation by direct discretization

```
f:=linsolve(a,h):
for `i` from 1 to twok+1
do
twotheta_f[`i`]:=twotheta_h[1]+(`i`-1)*deltatwotheta:
intensity_f[`i`]:=eval(f[`i`]);
end do:
p_h:=plot([twotheta_h[`ihh`],intensity_h[`ihh`],`ihh`=1..k],col-
or=red,style=LINE,thickness=2,axes=boxed,gridlines,labels=["2theta",""]):
p_g:=plot([twotheta_g[`igg`],intensity_g[`igg`],`igg`=1..k],
color=blue,style=LINE,thickness=2,axes=boxed,gridlines,
labels=["2theta",""]):
p_f:=plot([twotheta_f[`iff`],intensity_f[`iff`],`iff`=1..twok+1],
color=green,style=LINE,thickness=2,axes=boxed,gridlines,
labels=["2theta",""]):
display({p_h,p_g,p_f});
fd:= fopen("f",WRITE,TEXT):
for `i` from 1 to k
do
fprintf(fd,"%g %g\n",twotheta_f[`i`],intensity_f[`i`]):
end do:
fclose(fd):
```

Appendix 2

Fourier transform of true sample function procedure

```
f_GFF_freq:=proc(freq)
local arg_in,arg_sa;
arg_in:=(Pi*q_in)/(2*rho_in) + I *(Pi*Pi*freq)/rho_in;
arg_sa:=(Pi*q_sa)/(2*rho_sa) + I * (Pi*Pi*freq)/rho_sa;
(ampl_sa/ampl_in)*(rho_in/rho_sa)*cos(arg_in)/cos(arg_sa);
end:
```

Module of Fourier transform of true sample function procedure

```
FT_GFF_modul:=proc(freq)
local aux,bux,auxr_0,auxi_0;
aux:=evalc(Re(f_GFF_freq(freq))):
bux:=evalc(Im(f_GFF_freq(freq))):
aux:=aux*aux+bux*bux:
auxr_0:=evalc(Re(f_GFF_freq(0))):
auxi_0:=evalc(Im(f_GFF_freq(0))):
```

```
bux:=auxr_0*auxr_0+auxi_0*auxi_0:
sqrt(aux/bux):
end:
```

True sample function procedure

```
f_GFF_s:=proc(s)
local arg1,arg2;
arg1:=(Pi/2)*(rho_sa/rho_in);
arg2:=(rho_sa*s);
(2/Pi)*(ampl_sa/ampl_in)*
rho_in*cos(arg1)*cosh(arg2)/(cosh(2*arg2)+cos(2*arg1));
end:
```

Integral width of true sample function procedure

```
int_width_GFF:=proc(rho_in,rho_sa)
local arg;
arg:=Pi*rho_sa/rho_in;
Pi/(2*rho_sa*cos(arg/2))*(cos(arg)+1);
end:
```

Moment of zero order for experimental X-ray line profile procedure

```
mu_0_GFF:=proc(ampl,rho,q)
local arg;
arg:=(Pi*q)/(2*rho);
(ampl/2)*(Pi/rho)*(1/cos(arg));
end:
```

Moment of first order for experimental X-ray line profile procedure

```
mu_1_GFF:=proc(rho,q)
local arg;
arg:=(Pi*q)/(2*rho);
(Pi/(2*rho))*tan(arg);
end:
```

Moment of second order for experimental X-ray line profile procedure

```
mu_2_GFF:=proc(rho,q)
local arg;
arg:=(Pi*q)/(2*rho);
((Pi/(2*rho))^2)*(1./(cos(arg)^2)+tan(arg)^2);
end:
```

Experimental X-ray line profile procedure approximated by GFF distribution

```
exp_profile_GFF:=proc(s)
local arg_q,arg_rho;
arg_q:=q*s; arg_rho:=rho*s;
(ampl/2)*(cosh(arg_q)+sinh(arg_q))/cosh(arg_rho);
end:
```

Instrumental X-ray line profile procedure determined by GFF distribution

```
inst_profile_GFF:=proc(s)
local arg_q,arg_rho;
arg_q:=q_in*s; arg_rho:=rho_in*s;
(ampl_in/2)*(cosh(arg_q)+sinh(arg_q))/cosh(arg_rho);
end:
```

Fourier transform procedure for general relation of true sample function developed by Warren-Averbach theory

```
gen_function:=proc(freq)
exp(-beta_gen(fmin,fmax)*freq*freq-
gama_gen(fmin,fmax)*abs(freq));
end:
```

Procedure for experimental XRLP given by Voigt approximation

```
h_Voigt_function:=proc(s)
local arg1,arg2,arg3,arg4;
arg1:=(gama_h_c**2-s**2)/(gama_h_g**2);
arg2:=(gama_h_c-I*s)/gama_h_g;
arg3:=(gama_h_c+I*s)/gama_h_g;
arg4:=2.*gama_h_c*s/(gama_h_g**2);
amp_h/(sqrt(Pi)*gama_h_g)*exp(arg1)*
(Re(erfc(arg2))*cos(arg4)-Im(erfc(arg3))*sin(arg4));
end:
```

Procedure for instrumental XRLP given by Voigt approximation

```
g_Voigt_function:=proc(s)
local arg1,arg2,arg3,arg4;
arg1:=(gama_g_c**2-s**2)/(gama_g_g**2);
arg2:=(gama_g_c-I*s)/gama_g_g;
arg3:=(gama_g_c+I*s)/gama_g_g;
arg4:=2.*gama_g_c*s/(gama_g_g**2);
amp_g/(sqrt(Pi)*gama_g_g)*exp(arg1)*
(Re(erfc(arg2))*cos(arg4)-Im(erfc(arg3))*sin(arg4));
end:
```

Procedure for the true sample function calculated by Voigt approximation

```
f_Voigt_function:=proc(s)
local arg1,arg2,arg3,arg4;
arg1:=(gama_f_c**2-s**2)/(gama_f_g**2);
arg2:=(gama_f_c-I*s)/gama_f_g;
arg3:=(gama_f_c+I*s)/gama_f_g;
arg4:=2.*gama_f_c*s/(gama_f_g**2);
amp_h/amp_g/(sqrt(Pi)*gama_f_g)*exp(arg1)*(Re(erfc(arg2))*cos(arg4)-
Im(erfc(arg3))*sin(arg4));
end:
```

Acknowledgements

Financial support received from the European Union through the European Regional Development Fund, Project ID 1822/SMIS CSNR 48797 CETATEA, is gratefully acknowledged. In particular, one of the topics covered by the book *Fourier Transform of the Signals* will be a useful starting point in accomplishing one of its major objectives, energy recovery from ambient pollution. Additionally, the authors are grateful to the staff of Beijing Synchrotron Radiation Facilities for beam time and for their technical assistance in XRD measurements.

Author details

Adrian Bot[1], Nicolae Aldea[1*] and Florica Matei[2]

*Address all correspondence to: naldea@itim-cj.ro

1 National Institute for Research and Development of Isotopic and Molecular Technologies, Cluj-Napoca, Romania

2 University of Agricultural Sciences and Veterinary Medicine, Cluj-Napoca, Romania

References

[1] Kabanikin S. Inverse and Ill-Posed Problems: Theory and Applications. Berlin/Boston: Walter de Gruyter GmbH; 2011.

[2] Nedelcov I. P. Improper Problems in Computational Physics. Computer Physics Communications 1972; 4(2)157-163.

[3] Read R. J. Intensity Statistics in the Presence of Translational Noncrystallographic Symmetry. Acta Crystallographica Section D 2013; 69(2) 176-183.

[4] Kleefeld A., Cottbus B. T. U. Numerical Results for Linear Fredholm Integral Equations of the First Kind over Surfaces in 3D. International Journal of Computer Mathematics 2010; 1(1) 1-16.

[5] Naumova V., Pereverzyev S. V. Multi-penalty Regularization with a Componentwise Penalization. Inverse Problems 2013; 29(7) 1-16.

[6] Kabanikin S., editor. Regularization Theory for Ill-posed Problems: Selected Topics. Inverse and Ill Posed Problems Series 58. Berlin/Boston: Walter de Gruyter GmbH; 2013.

[7] Beniaminy I., Deutsch M. A Spline Based Method for Experimental Data Deconvolution. Computer Physics Communications 1980; 21(2): 271-277.

[8] Fredrikze H., Verkerk P. Comment on "A spline based method for experimental data deconvolution". Computer Physics Communications 1981; 24(1), 5-7.

[9] Aldea N., Tiusan C., Zapotinschi R. A New Approach Used to Evaluation the Crystallite Size of Supported Metal Catalysts by Single X-Ray Profile Fourier Transform Implemented on Maple V. In: Borcherds P., Bubak M., Maksymowicz A., editors. 8th Joint EPS-APS International Conference on Physics Computing, PC '96 17-21 September 1996, Krakow: Academic Computer Centre CYFRONET-KRAKOW; 1996.

[10] Balzar D., Ledbetter H. J. Voigt-Function Modeling in Fourier-Analysis of Size and X-ray Diffraction Peaks-Peaks. Journal of Applied Crystallography 1993; 26(1) 97-103.

[11] Aldea N., Gluhoi A., Marginean P., Cosma C., Yaning X. Extended X-Ray Absorption Fine Structure and X-Ray Diffraction Studies on Supported Nickel Catalysts. Spectrochimica Acta Part 2000; 55(7) 997-1008.

[12] Aldea N., Barz B., Pintea S., Matei F. Theoretical Approach Regarding Nanometrology of the Metal Nanoclusters Used in Heterogeneous Catalysis by Powder X-Ray Diffraction Method. Journal of Optoelectronics and Advanced Materials 2007; 9(10) 3293-3296.

[13] Gradstein I. S., Rijik L. M. Tables of Integrals, Sums, Series, and Products. Moscow: Fizmatgiz; 1962.

[14] Balzar D., Popa N. C. Crystallite Size and Residual Strain/Stress Modeling in Rietveld Refine. In: Meittemeijer E. J., Scardi P., editors. Diffraction Analysis of the Microstructure of Materials. Berlin Heidelberg New York: Springer-Verlag; 2003, pp. 125-144.

[15] Lazar M., Valer A., Pintea S., Barz B., Ducu C., Malinovschi V., Xie Yaning Aldea N. Preparation and Structural Characterization by XRD and XAS of the Supported Gold Catalysts. Journal of Optoelectronics and Advanced Materials 2008; 10(9) 2244-2251.

[16] Walker J. S. Fast Fourier Transform. 2nd ed. New York, London, Tokyo: Boca Raton CRC; 1997.

[17] Balzar D. BREADTH- a Program for Analyzing Diffraction Line Broadening. Journal of Applied Crystallography 1995; 28(2) 244-245.

[18] Rednic V., Aldea N., Marginean P., Rada M., Bot A., Zhonghua W., Zhang J., Matei F. Heat Treatment Influence on the Structural Properties of Supported Ni Nanoclusters. Metals and Materials International 2014; 20(4) 641-646.

5

Efficient FFT-based Algorithms for Multipath Interference Mitigation in GNSS

Renbiao Wu, Qiongqiong Jia, Wenyi Wang and
Jie Li

Additional information is available at the end of the chapter

1. Introduction

GNSS (Global Navigation Satellite System) has been found application in many areas. In some cases, the performance requirements for GNSS are very high. There are many error sources that would degrade the positioning performance of GNSS, e.g., clock errors, ephemeris errors, tropospheric propagation delay, and multipath. Many positioning errors mentioned above are constant for all GNSS receivers in a given small area and can be removed or reduced by using the popular differential technique. However, due to the geographical position difference between the reference station and the receiver, the multipath environment of receivers, such as amplitudes and number of multipath signals, is totally different with that of the reference station. Thus differential technique can not eliminate the multipath error. Many studies have shown that the multipath interference will lead to a position error around several meters which endangers the reliability and accuracy of GNSS. Therefore, multipath interference mitigation has been a hot topic in the field of satellite navigation receiver design.

Multipath interferences are the signals reflected by the objects around the GNSS receiver. Then the multipath interference and the LOS (line of sight) signal are simultaneously received by antenna which brings a phase distortion in the tracking loops of receivers. Finally, the phase distortion results in the tracking and positioning error.

There have been many studies about the effect of multipath interference. Kalyanaraman et al. in [1] analyzed the multipath effects on code tracking loop. Kos et al. [2] provided a detailed analysis of the multipath effects on the positioning error. Main multipath interference mitigation techniques are based on antenna technique and signal processing algorithms.

By mounting the antenna in a well-designed place based on the multipath environment, Maqsood et al. in [3] compared the performances of multipath interference mitigation for different antennas. Ray et al. in [4] proposed a multipath mitigation algorithm based on an antenna array. A design principle of antenna was proposed by Alfred et al. in [5] for satellite navigation landing system. The multipath mitigation technique based on the special antenna can only suppress the multipath signal coming from the ground below the antenna. However, it is useless for the multipath interference signal reflected by the objects above the antenna.

Popular signal processing techniques are the narrow correlator and MEDLL (Multipath Estimated Delay Locked Loop). Narrow correlator is adopted in many GNSS receivers which suppresses multipath interference by reducing the early-late correlator spacing. The signal model for the narrow correlator is provided by Michael et al. in [6]. Cannon et al. in [7] analyzed the performance of narrow correlator in the satellite navigation system. The performance of narrow correlator can be improved by decreasing the early-late correlator spacing. However, the narrow correlator assumes that the bandwidth of the received signal is infinite which is invalid in the practical applications. Thus, when the early-late correlation spacing is less than the reciprocal of the channel bandwidth, the tracking error cannot be further decreased by reducing the early-late correlator spacing. MEDLL is a multipath interference mitigation algorithm based on the statistical theory as in [8], which estimates the time delays via the maximum likelihood criterion. Therefore, the complexity of MEDLL is much higher than narrow correlator.

This chapter firstly presents a new acquisition algorithm for GNSS. Then two multipath interference mitigation algorithms based on the DPE (Decoupled Parameter Estimation) parameter estimation algorithms are presented. The FFT algorithm is utilized to reduce the computational complexity in all proposed algorithms. Numerical results are provided to demonstrate the performances of the proposed algorithms. The remainder of this chapter is arranged as follows. Section 2 presents the new acquisition algorithm. Two multipath interference mitigation algorithms are separately presented in Section 3. Section 4 concludes the chapter.

2. A novel GNSS acquisition algorithm

2.1. Data model and problem formulation

GNSS signal is composed of three parts, the carrier, the ranging codes and the data codes. In order to facilitate the presentation, GPS is taken as an example. However, the proposed algorithm is also suitable for other GNSS system. The carrier of GPS (Global Navigation System) consists of the three different frequency bands, i.e., L1 (1575.42 MHz), L2 (1227.6 MHz) and L5 (1176.45MHz), see [9-10]. The ranging codes are pseudo random noise (PRN) codes including C/A code, P code and M code which are known in advance. Data codes or the navigation data are binary codes containing the satellite ephemeris, satellite status, clock system, the orbit perturbation correction of the satellite clock motion state and the atmospheric refraction correction, etc..

Suppose that one receiver obtained the signals from P satellites and the received data can be expressed as

$$y(t) = \sum_{p=1}^{P} \alpha_p d_p(t - \tau_p) c_p(t - \tau_p) e^{j\omega_{dp}(t-\tau_p)} + e(t) \tag{1}$$

where $d_p(t)$ represents the navigation data of the pth satellite, $c_p(t)$ is the C/A code of the pth satellite, $e(t)$ is the thermal noise, α_p, τ_p and ω_{dp} denote the amplitude, time delay and Doppler frequency of the pth satellite, respectively. It should be noted that only the L1 signal is considered in this chapter.

After A/D conversion, the data can be written as

$$y(n) = \sum_{p=1}^{P} \alpha_p d_p(n - \tau_p) c_p(n - \tau_p) e^{j\omega_{dp}(n-\tau_p)} + e(n) \tag{2}$$

The conventional acquisition procedure is a two-dimensional searching algorithm, which is time consuming even when using parallel searching algorithm. Moreover, the frequency resolution of the conventional acquisition algorithm cannot satisfy the requirements of the tracking loop. Hence, a more accurate frequency estimation is required before tracking in the conventional receiver.

Therefore, a new GNSS acquisition algorithm is proposed in this section. The Doppler frequency is firstly estimated. After that, the initial code phase is then obtained via NLS (Nonlinear Least Square) fitting. Compared with the conventional acquisition algorithm, the proposed algorithm can obtain a comparative performance with a lower computational complexity.

2.2. Principle of the novel acquisition algorithm

It can be noted from the data model in section 2.1 that there are three unknown parameters in the acquisition process, which are the PRN index of the pth satellite, the corresponding Doppler frequency ω_{dp} and its initial code phase τ_p.

2.2.1. Doppler frequency estimation

Due to the navigation data and C/A code are ±1 in GPS system, it can be seen from equation (2) that the Fourier spectrum of each satellite contains multiple frequency components. Therefore, the Doppler frequency cannot be extracted from the spectrum directly. To eliminate the influence on frequency estimation brought by the navigation data and C/A code, we square equation (2) as in [11]

$$y^2(n) = \sum_{p=1}^{P} \alpha_p^2 d_p^2(n-\tau_p) c_p^2(n-\tau_p) e^{j2\omega_{dp}(n-\tau_p)} + 2e(n)\sum_{p=1}^{P} \alpha_p d_p(n-\tau_p) c_p(n-\tau_p) e^{j\omega_{dp}(n-\tau_p)}$$

$$+2\sum_{p=1}^{P-1}\sum_{r=p+1}^{P} \alpha_p \alpha_r d_p(n-\tau_p) d_r(n-\tau_r) c_p(n-\tau_p) c_r(n-\tau_r) e^{j[\omega_{dp}(n-\tau_p)+\omega_{dr}(n-\tau_r)]} + e^2(n) \tag{3}$$

Since the value of navigation data and C/A code is ±1 in GPS system, then the above equation can be simplified as

$$y^2(n) = \sum_{p=1}^{P} \alpha_p^2 e^{j2\omega_{dp}(n-\tau_p)} + e_1(n) \tag{4}$$

where,

$$e_1(n) = 2\sum_{p=1}^{P-1}\sum_{r=p+1}^{P} \alpha_p \alpha_r d_p(n-\tau_p) d_r(n-\tau_r) c_p(n-\tau_p) c_r(n-\tau_r) e^{j[\omega_{dp}(n-\tau_p)+\omega_{dr}(n-\tau_r)]}$$

$$+2e(n)\sum_{p=1}^{P} \alpha_p d_p(n-\tau_p) c_p(n-\tau_p) e^{j\omega_{dp}(n-\tau_p)} + e^2(n) \tag{5}$$

It can be seen from equation (4) that the spectrum of each satellite only contains a single dominant frequency component corresponding to the Doppler frequency. For the reason that not only the correlation of C/A code of different satellites is very small but also the signal and noise are uncorrelated, it is obviously that in equation (5) the product terms of different C/A codes as well as the product of noise and signal are close to zero. Fourier analysis method can be directly used to estimate the signal frequency.

The Fourier transform of equation (4) can be expressed as

$$F(\omega) = \left| \sum_{n=-N/2}^{N/2-1} y^2(n) e^{-j\omega n} \right| \tag{6}$$

where $\omega = 2\omega_{dp}$ can be obtained based on the locations of the first p dominant peaks of $F(\omega)$. A more accurate estimation result can be achieved by using FFT (Fast Fourier Transform) padding with zeros.

The estimated value of $\hat{\omega}$ is in the range of $[-\pi, \pi]$, thus from equation (6) we know that the estimated value of $\hat{\omega}_{dp}$ $[-\pi/2, \pi/2]$. As ω_{dp} $[-\pi, \pi]$, the estimated frequency could be $\hat{\omega}_{dp}$ or $\hat{\omega}_{dp} + \pi$. Therefore, the frequency ambiguity will lead to a mistake. Due to the IF (Intermediate Frequency) of the satellite signals is definitely known after down conversion and the range of Doppler shift is between ±10kHz [12], the difference between true frequency and the ambiguity

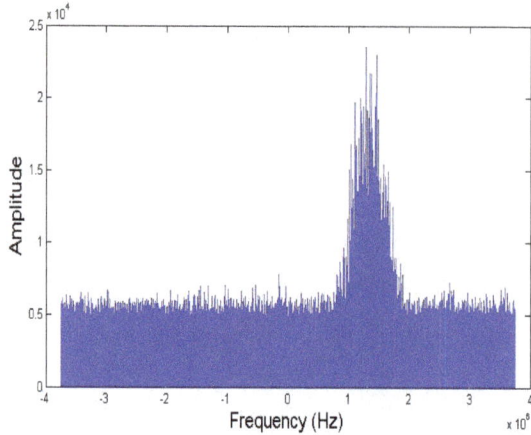

(a) before square operation

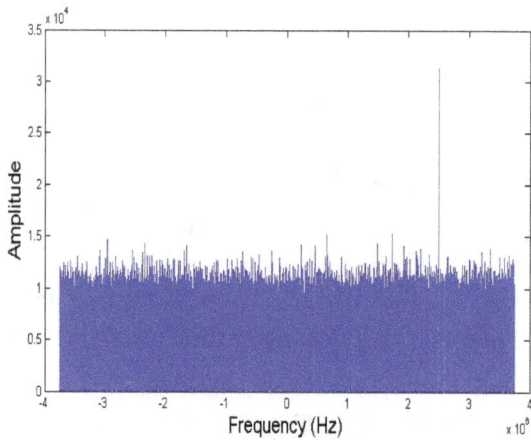

(b) after square operation

Figure 1. Frequency spectrum before and after square operation

one is generally greater than the Doppler shift. Therefore, ambiguity frequency can be excluded by determining whether the frequency is in the range of the Doppler shift.

The frequency spectrum of the received data before and after the square operation is shown in Figure 1. It can be noted that before square operation, there are multiple dominant frequency components. Thus the Doppler frequency cannot be extracted directly. However, after being squared, there is only one dominant Doppler frequency present in the spectrum, so an accurate Doppler frequency can be estimated.

2.2.2. Code phase estimation and matching

After obtaining the Doppler frequency, the initial phases and the PRN index of the C/A code from all satellites can be estimated by using its autocorrelation characteristic. Let $s_p(n)=d_p(n)c_p(n)e^{j\omega_{dp}n}$, then equation (2) can be expressed as

$$y(n) = \sum_{p=1}^{P} \alpha_p s_p (n - \tau_p) + e(n) \tag{7}$$

As the Doppler frequency $\hat{\omega}_{dq}$ has been estimated, $\hat{\omega}_{dq}$ can be used to take the place of ω_{dq}. Then the signal sent by the qth satellite can be reconstructed as

$$\hat{s}_q(n) = d_q(n)c_q(n)e^{j\hat{\omega}_{dq}n} \tag{8}$$

Suppose the time delay from the qth satellite to the receiver is τ_q, then the received signal can be written as

$$y_q(n) = \alpha_q \hat{s}_q(n - \tau_q) + e_2(n) \tag{9}$$

where $e_2(n) = \sum_{\substack{p=1 \\ p \neq q}}^{P} \alpha_p s_p (n - \tau_p) + \alpha_q s(n - \tau_q) - \alpha_q \hat{s}(n - \tau_q) + e(n)$. The DFT (Discrete Fourier Transform) of equation (9) can be written as

$$y_f(k) = \alpha_q \hat{s}_{fq}(k)e^{j\omega_q k} + e_{f2}(k) \tag{10}$$

where $y_f(k)$, $\hat{s}_{fq}(k)$ and $e_{f2}(k)$ are the DFT of $y(n)$, $\hat{s}_q(n)$ and $e_2(n)$ respectively, $\omega_q = -2\pi\tau_q f_s / N$ with f_s is the sampling frequency. Note, here x_f is used to represent the DFT of x. Hence, the estimation of τ_q is transformed to the estimation of ω_q.

To estimate ω_q, we define the following NLS cost function [12-15]

$$z_f(\hat{\omega}_q) = \sum_{k=-N/2}^{N/2-1} \left| y_f(k) - \hat{s}_{fq}(k)\alpha_q e^{j\omega_q k} \right|^2 \tag{11}$$

Let

$$\hat{S}_{fq} = diag\left\{ \hat{s}_{fq}(-N/2), \hat{s}_{fq}(-N/2+1), ..., \hat{s}_{fq}(N/2-1) \right\} \tag{12}$$

$$\mathbf{a}(\omega_q) = \left[e^{j\omega_q(-N/2)}, e^{j\omega_q(-N/2+1)}, ..., e^{j\omega_q(N/2-1)} \right]^T \tag{13}$$

$$\mathbf{y}_f = \left[y_f(-N/2), y_f(-N/2+1), ..., y_f(N/2-1) \right]^T \tag{14}$$

where $(\quad)^T$ denotes the transpose operation. Then the cost function in equation (11) can be transformed as

$$Z\left(\hat{\omega}_q\right) = \left\| \mathbf{y}_f - \alpha_q \hat{\mathbf{S}}_{fq} \mathbf{a}(\omega_q) \right\|^2 \tag{15}$$

where $\| \quad \|$ denotes the Euclidean norm. Minimizing $Z(\hat{\omega}_q)$ with respect to ω_q yields the estimated $\hat{\omega}_q$ as

$$\hat{\omega}_q = \arg \max_{\omega_q} \left| \mathbf{a}^H(\omega_q) \hat{\mathbf{S}}_{fq}^H \mathbf{y}_f \right|^2 \tag{16}$$

The time delay estimation can be further obtained by

$$\hat{\tau}_q = -\hat{\omega}_q N / (2\pi f_s) \tag{17}$$

Equation (16) can be solved by using FFT algorithm with a low computation burden. Since the PRN index of the received signal is unknown, consequently, it is not feasible to reconstruct $\hat{s}_q(n)$ directly by equation (8). Therefore, we should reconstruct signal of each satellite with the estimated Doppler frequency $\hat{\omega}_q$. Then the cross correlation of the reconstructed signal and the received one can be calculated. Furthermore, the corresponding time delay can be estimated from equation (16).

It is well-known that the cross correlation coefficients of different C/A codes are very small. In addition, the signal and noise are uncorrelated. Thus Fourier analysis results of the above mentioned component are close to zero. The maximum correlation value can only be obtained in the case that the C/A code of the reconstructed signal is the same as the received one. Therefore, the PRN index can be obtained. Furthermore, the code delay can be estimated by the location of the maximum correlation value. Then, we obtained all the unknown parameters.

2.3. Comparison of acquisition algorithms

Suppose the searching time for Doppler frequency estimation is Q in the conventional acquisition algorithm, P is the number of satellite received by the receiver. In GPS, the typical values of Q and P are 21 and 4-8, respectively, as in [9, 11]. The total number of satellites M is 32. Since Doppler frequency and code delay are unknown in the conventional acquisition algorithm, Q frequency points should be searched over certain frequency range. What's more, M satellites are also to be searched for each of the Q frequency points. While in the proposed algorithm, the Doppler frequencies of P satellites have been estimated, only P accurate frequencies need to be searched.

Figure 2 compares the searching process of the proposed algorithm and the conventional one. It can be seen that the computation complexity of the proposed algorithm is significantly lower than that of the conventional algorithm in that the typical value of Q is obviously greater than P.

Comparison of computation burden between new acquisition algorithm and conventional acquisition algorithm are shown Table 1. It can be clearly seen from Table 1 that, since Q is

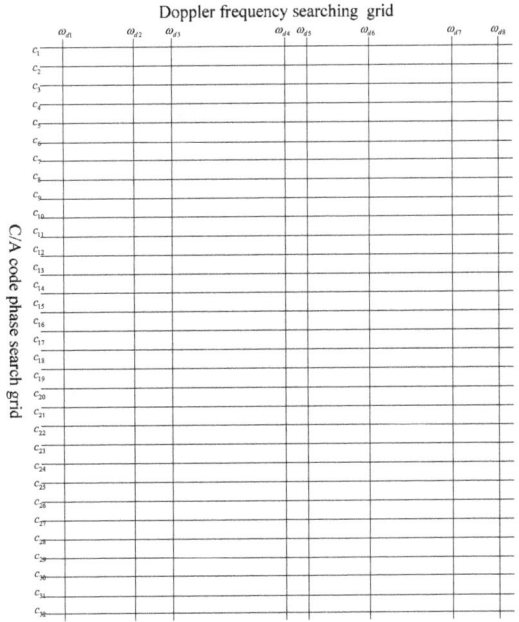

(a) searching process of new algorithm

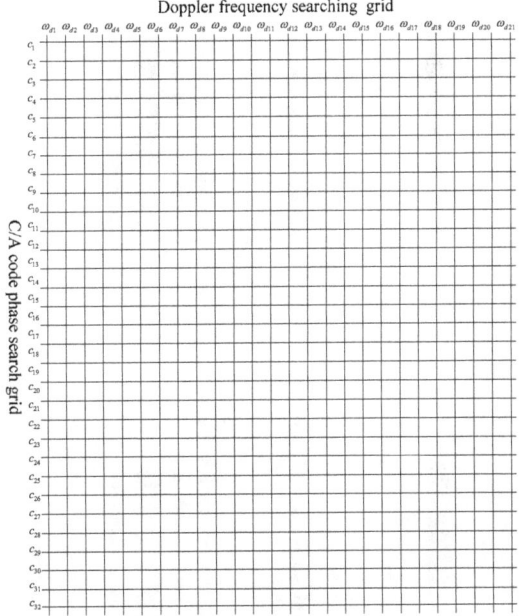

(b) searching process of conventional algorithm

Figure 2. Comparison of searching process between new algorithm and conventional algorithm

always greater than P, the computation complexity of the proposed algorithm is lower than the conventional acquisition procedure. In practices, only 4 satellites are enough to determine the user position, which is less than 8, hence the computation burden should be further decreased. As the number of satellite is equal to 4, only half of the original computation complexity is required.

	Multiplication times	Addition times	FFT times	Time Complexity
conventional method	7MQ+4M	2MQ+3M	2MQ	O(MQ)
new method	6MP+4P	3MP+7P	M (P+1) +1	O(MP)

Table 1. Comparison of computation burden between new algorithm and conventional one

2.4. Numerical results

To verify the performance of the proposed algorithm, GPS signals from 8 satellites with PRN indexes 1, 2, 13, 20, 22, 24, 25 and 27 are simulated. Considering the IF is 1.25MHz, the sampling rate is 7.5MHz, the pre integration time is 1ms and the signal to noise ratio (SNR) is -20dB. The frequency searching step of the conventional acquisition algorithm is 1kHz. The acquisition results are shown in Figure 3.

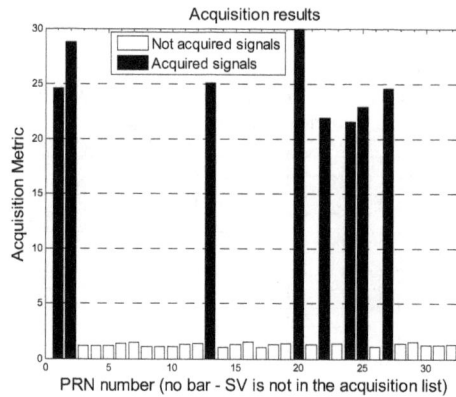

(a)   conventional acquisition algorithm

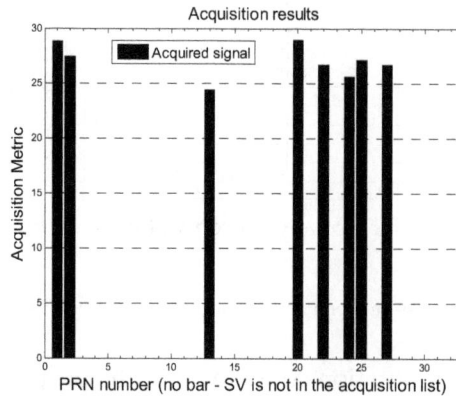

(b) proposed acquisition algorithm

Figure 3. Compassion of the acquisition results

In Figure 3, the horizontal axis stands for the PRN index of the satellites and the vertical axis denotes the acquisition metric. Since the proposed algorithm only calculates the correlation of the received satellite signal and the local reference signal, correlation results of the other satellite are not shown in Figure 3(b). It can be noted that acquisition results are identical, and the peak value of the proposed algorithm is higher than the conventional one. Therefore, the proposed algorithm has a comparative performance with the conventional one.

3. Two novel methods for multipath mitigation

3.1. Multipath data model

The term multipath is derived from the fact that a signal transmitted from a GNSS satellite can follow a 'multiple' number of propagation 'paths' to the receiving antenna. This is possible because the signal can be reflected back to the antenna from surrounding objects, including the earth's surface.

Suppose the signal transmitted by GNSS satellites can be written as

$$x(t) = d(t)c(t)e^{j\omega_c t} \tag{18}$$

where $d(t)$ is the navigation data, $c(t)$ is the C/A code in GPS. Here, we take the two path signal model as an example, which can be generalized to multiple reflection path signals directly. The received signal including the LOS signal with a single reflection can be written as

$$\bar{y}(t) = \sum_{p=1}^{2} \alpha'_p d(t - \tau_p) c(t - \tau_p) e^{j\omega_c(t - \tau_p)} + e(t) \tag{19}$$

where α_1', τ_1 is amplitude and code delay of the LOS signal, α_2', τ_2 is amplitude and code delay of the reflect signal, $e(t)$ is the thermal noise. Equation (19) can also be deployed as

$$\begin{aligned}
\bar{y}(t) &= \sum_{p=1}^{2} \alpha'_p d(t - \tau_p) c(t - \tau_p) e^{j\omega_c(t - \tau_p)} + e(t) \\
&= \left(\alpha_1' d(t - \tau_1) c(t - \tau_1) e^{-j\omega_c \tau_1} + \alpha_2' d(t - \tau_2) c(t - \tau_2) e^{-j\omega_c \tau_2} \right) e^{j\omega_c t} + e(t)
\end{aligned} \tag{20}$$

To obtain the Doppler frequency of the LOS signal, we use the following relation

$$\phi = \omega_c \tau_1 = \omega_c \frac{R_1(t)}{c} \tag{21}$$

Assume the radial velocity of the receiver relative to the satellite is v_1, then the range between them can be given by

$$R_1(t) = R_0 + v_1 t \tag{22}$$

where $\mathbf{R}_0$ is the initial range between receiver and satellite, then equation (21) can be further given by

$$\phi = \omega_c \tau_1 = \omega_c \frac{R_0 + v_1 t}{c} = 2\pi \left(\frac{R_0}{\lambda} + \frac{v_1 t}{\lambda} \right) \tag{23}$$

As the relative radial velocity is constant, the Doppler frequency can be obtained by

$$\omega_{d1} = \frac{d\phi}{dt} = \frac{v_1}{\lambda} \tag{24}$$

The propagation range of the multipath signal can be written as

$$R_2(t) = R_0 + v_1 t + \Delta R(t) \tag{25}$$

where $\Delta \mathbf{R}(t)$ is the propagation range difference between the LOS signal and the reflected one. Since satellites are far away from the receiver, the range difference $\Delta \mathbf{R}(t)$ can be considered as constant in the short integration time. Then equation (21) can be represented as

$$\omega_c \tau_p = \omega_c \frac{R_0 + v_1 t}{c} = 2\pi \left(\frac{R_0}{\lambda} + \frac{v_1 t}{\lambda} + \frac{\Delta R}{\lambda} \right) \tag{26}$$

Further simplify of equation (20) we can get

$$\bar{y}(t) = \left(\alpha_1' d(t - \tau_1) c(t - \tau_1) e^{-j\varphi_1} + \alpha_2' d(t - \tau_2) c(t - \tau_2) e^{-j(\varphi_1 + \Delta\varphi)} \right) e^{j2\pi\omega_d' t} + e(t) \tag{27}$$

where $\omega_d' = \omega_{d1} + \omega_c$ is the Doppler frequency of the received signal, $\varphi_1 = \mathbf{R}_0 / \lambda$ and $\varphi_2 = \varphi_1 + \Delta\varphi$ are the phase of the LOS signal and the reflect one respectively, with $\Delta\varphi$ an extra phase caused by the extra propagation range $\Delta \mathbf{R}$. Accordingly, the LOS signal and reflect signal share the same Doppler shift.

After A/D conversion, the transformed digital signal can be written as

$$\bar{y}(n) = \sum_{p=1}^{2} \alpha'_p d(n - \tau_p) c(n - \tau_p) e^{-j\varphi_p} e^{j\omega'_d n} + e(n) \tag{28}$$

Assume the Doppler frequency has been estimated accurately, then the complex phase $e^{j\omega'_d n}$ can be compensated after down-conversion. Hence, equation (28) can be written as

$$y(n) = \sum_{p=1}^{2} \alpha'_p d(n - \tau_p) c(n - \tau_p) e^{-j\varphi_p} + \acute{e}(n) \tag{29}$$

where $\acute{e}(n)$ is equal to $e(n) e^{-j\omega'_d n}$ that share the same statistics characteristic with $e(n)$. So we still use $e(n)$ to represent the noise data.

The in-phase component of equation (29) can be given by

$$y_1(n) = \sum_{p=1}^{2} \alpha'_p d(n - \tau_p) c(n - \tau_p) \cos(\varphi_p - \varphi_c) + e(n) \tag{30}$$

where φ_c is the synthetic phase of the LOS signal and the reflect one. Assume d is the early-late correlator spacing in the classical DLL, the estimated code delay of the direct signal is $\hat{\tau}_1$. Then the early and late code can be written as

$$s_E(t) = y_I(t - \hat{\tau}_1 - d/2) \tag{31}$$

$$s_L(t) = y_I(t - \hat{\tau}_1 + d/2) \tag{32}$$

In classical DLL, the in-phase early and late correlation value is given as

$$R_E(\varepsilon) = \sum_{p=1}^{2} \alpha_p R(\varepsilon + \Delta\tau_p - d/2) \cos(\varphi_p - \varphi_c) \tag{33}$$

$$R_L(\varepsilon) = \sum_{p=1}^{2} \alpha_p R(\varepsilon + \Delta\tau_p + d/2) \cos(\varphi_p - \varphi_c) \tag{34}$$

where $\varepsilon = \tau_1 - \hat{\tau}_1$ is the estimation error of the LOS signal, namely tracking error. $\Delta\tau = \tau_2 - \tau_1$ is the relative delay of the multipath signal to the LOS. The discrimination function of the classical DLL (Delay Locked Loop) can be written as

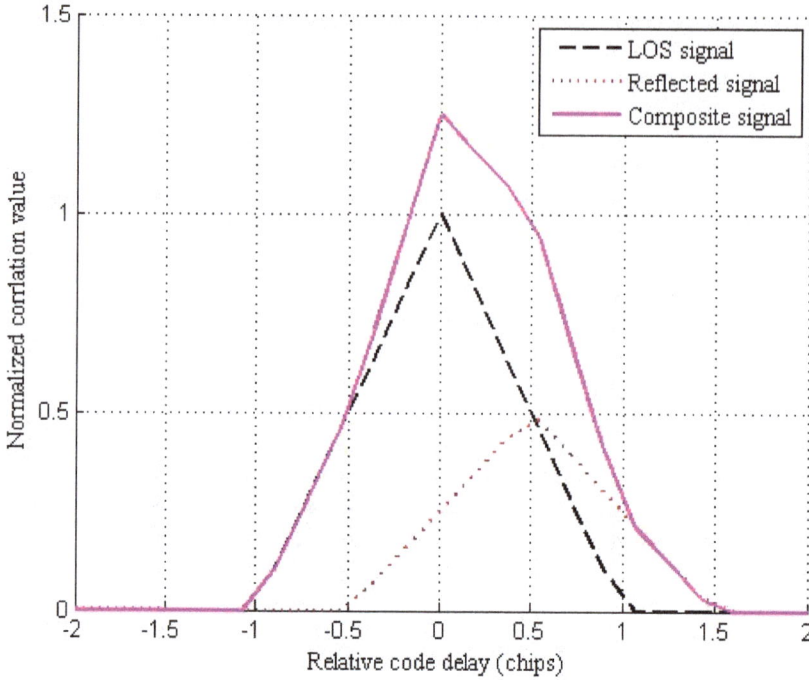

Figure 4. Correlation peak when multipath is present

$$D(\varepsilon) = R_E(\varepsilon) - R_L(\varepsilon)$$

$$= \sum_{p=1}^{2} \alpha_p \left[R(\varepsilon + \Delta\tau_p - d/2) - R(\varepsilon + \Delta\tau_p + d/2) \right] \cos(\varphi_p - \varphi_c) \qquad (35)$$

In GNSS navigation, the receiver is concerned to maximize the correlation function between the received and locally generated signals. This can be accomplished by determining the locations of the zero output of the discriminator which corresponds to the maximum of the correlation function. Here, the early-late correlator is used to determine the position of this zero. However, the presence of multipath introduces some bias in the position of the first arrival peak and has an impact in the user's position, which can be clearly seen in Figure 4. Then the classical DLL failed to cope with multipath propagation, see [7-8].

3.2. Code delay estimation

3.2.1. Code delay estimation in the correlation domain via NLS

In this subsection, a code delay estimation algorithm based on DPE in correlation domain is proposed. To deploy the proposed algorithm, we combine the complex constant phase $e^{-j\varphi_p}$ and the real amplitude α'_p in equation (29) together as a new complex variable α''_p. Then equation (29) can be written as

$$y(n) = \sum_{p=1}^{2} \alpha''_p d(n - \tau_p) c(n - \tau_p) + e(n) \tag{36}$$

The navigation data cycle is much longer than that of the C/A code, but only one cycle of the C/A code is used in the code delay estimation. Consequently, the navigation data jump can be neglected in the signal reconstruction. Then the navigation data +1 or -1 is written into α''_p and denoted as α_p. Therefore equation (36) can be written as

$$y(n) = \sum_{p=1}^{2} \alpha_p c(n - \tau_p) + e(n) \tag{37}$$

To simplify the following expression, $s()$ is introduced to represent $c()$. Then equation (37) can be represented as

$$y(n) = \sum_{p=1}^{2} \alpha_p s(n - \tau_p) + e(n) \tag{38}$$

Multipath interference mitigation based on code delay estimation focus on the estimation of the LOS delay τ_1 and the reflect delay τ_2, where the estimated result can be used to migrate the multipath. After acquisition, the correlation between the received satellite signal and the local reference signal is used to estimate the parameter of both the LOS signal and the reflected one. The proposed code delay estimation algorithm can take the place of the classical DLL, which can be described in details as follows.

To obtain the initial code delay, the correlation function between the received satellite signal and the reference signal is represented as

$$r(n) = \mathrm{corr}\big(s(n), y(n)\big) \tag{39}$$

where corr() represents the cross correlation operation. Since the noise is uncorrelated with the signal, equation (39) can be further expressed as

$$r(n) = \sum_{p=1}^{2} \alpha_p r^s(n - \tau_p) + w(n) \tag{40}$$

where $r^s(n)$ is the autocorrelation of $s(n)$, $w(n)=\mathrm{corr}(s(n), e(n))$ can still be thought as noise that sharing the same statistic characteristics with $e(n)$. Apply DFT to equation (40)

$$p_f(k) = r_f^s(k)\sum_{p=1}^{2}\alpha_p e^{j\omega_p k} + w_f(k) \tag{41}$$

where $p_f(k)$, $r_f^s(k)$, $w_f(k)$ are the DFT of $p(n)$, $r^s(n)$ and $w(n)$ respectively, $\omega_p = -2\pi\tau_p/NT_s$ with T_s is the sampling interval.

Define a NLS cost function as

$$C_1\left(\{\omega_p,\alpha_p\}_{p=1}^{2}\right) = \sum_{k=-N/2}^{N/2}\left\|p_f(k) - r_f^s(k)\sum_{p=1}^{2}\alpha_p e^{j\omega_p k}\right\|^2 \tag{42}$$

The unknown parameters $\{\omega_p, \alpha_p\}_{p=1}^{2}$ can be obtained by minimizing the cost function $C_1(\{\omega_p, \alpha_p\}_{p=1}^{2})$. Whereas, searching over a multidimensional space to solve the NLS problem requires higher computational complexity. To reduce the complexity, the Weighted Fourier Transform and RELAXation (WRELAX) algorithm (see [12]) is utilized here to solve the NLS problem. Let

$$\mathbf{a}(\omega_p) = \left[e^{j\omega_p(-N/2)}, e^{j\omega_p(-N/2+1)}, ..., e^{j\omega_p(N/2-1)}\right]^{\mathrm{T}} \tag{43}$$

$$\mathbf{p}_f = \left[p_f(-N/2), p_f(-N/2+1), ..., p_f(N/2-1)\right]^{\mathrm{T}} \tag{44}$$

$$\mathbf{R}_f = diag\left\{r_f^s(-N/2), r_f^s(-N/2+1), ..., r_f^s(N/2-1)\right\} \tag{45}$$

Consequently the cost function in equation (42) can be rewritten as

$$C_1(\{\omega_p,\alpha_p\}_{p=1}^{2}) = \left\|\mathbf{p}_f - \sum_{p=1}^{2}\alpha_p\mathbf{R}_f\mathbf{a}(\omega_p)\right\|^2 \tag{46}$$

Further, we assume $\{\hat{\omega}_q, \hat{\alpha}_q\}_{q=1,q\neq p}^{2}$ have been estimated, then

$$\mathbf{p}_{fp} = \mathbf{p}_f - \sum_{\substack{q=1 \\ q\neq p}}^{2}\hat{\alpha}_q[\mathbf{R}_f\mathbf{a}(\hat{\omega}_p)] \tag{47}$$

Substitute equation (47) into equation (46) we have

$$C_2(\{\omega_p, \alpha_p\}_{p=1}^2) = \left\| \mathbf{p}_{fp} - \alpha_p \mathbf{R}_f \mathbf{a}(\omega_p) \right\|^2 \tag{48}$$

By minimizing the cost function C_2, estimation of $\hat{\omega}_p$ and $\hat{\alpha}_p$ can be obtained from

$$\hat{\omega}_p = \arg\max_{\omega_p} \left| \mathbf{a}^H(\omega_p)\left(\mathbf{R}_f^* \mathbf{p}_{fp}\right) \right|^2 \tag{49}$$

and

$$\hat{\alpha}_p = \left. \frac{\mathbf{a}^H(\omega_p)\left(\mathbf{R}_f^* \mathbf{p}_{fp}\right)}{\left\| \mathbf{R}_f \right\|_F^2} \right|_{\omega_p = \hat{\omega}_p} \tag{50}$$

where $\|\ \ \|_F$ represents the Frobenius norm. Hence, $\hat{\omega}_p$ can be obtained as the location of the periodogram $|\mathbf{a}^H(\omega_p)(\mathbf{R}_f \mathbf{p}_{fp})|^2$, which can be efficiently computed by using FFT to the data sequence $\mathbf{R}_f \mathbf{p}_{fp}$ padded with zeros. (Note the padding with zeros is necessary to determine $\hat{\omega}_p$ with more accuracy.) Then $\hat{\alpha}_p$ is easily computed from the complex height of the peak of $\dfrac{\mathbf{a}^H(\hat{\omega}_p)(\mathbf{R}_f \mathbf{p}_{fp})}{\| \mathbf{R}_f \|_F^2}$.

With the above simple preparations, we now proceed to present the correlation domain algorithm for the minimization of the NLS cost function. The proposed algorithm comprises the following steps.

Step 1. Assume $p=1$. Obtain $\{\hat{\omega}_1, \hat{\alpha}_1\}$ from equation (49) and equation (50).

Step 2. Assume $p=2$ (the LOS signal and one reflected path). Compute $\mathbf{p}_{f2}$ with (47) by using $\{\hat{\omega}_1, \hat{\alpha}_1\}$ obtained in Step (1). Obtain $\{\hat{\omega}_2, \hat{\alpha}_2\}$ from $\mathbf{p}_{f2}$ described above.

Next, compute $\mathbf{p}_{f1}$ with (47) by using $\{\hat{\omega}_2, \hat{\alpha}_2\}$ and redetermine $\{\hat{\omega}_1, \hat{\alpha}_1\}$ from $\mathbf{p}_{f1}$ as in equation (49) and equation (50) above.

Iterate the previous two substeps until "practical convergence" is achieved then we can obtain $\{\hat{\omega}_p, \hat{\alpha}_p\}_{p=1}^2$. Furthermore, by using $\hat{\omega}_p = -2\pi\hat{\tau}_p / NT_s$ the code delay $\hat{\tau}_p$, $p=1, 2$ can be obtained.

From the previous description one can find that the proposed algorithm can be implemented by simply FFT operation which leads to a less computation load.

The diagram of the novel code delay estimation algorithm is shown in Figure 5.

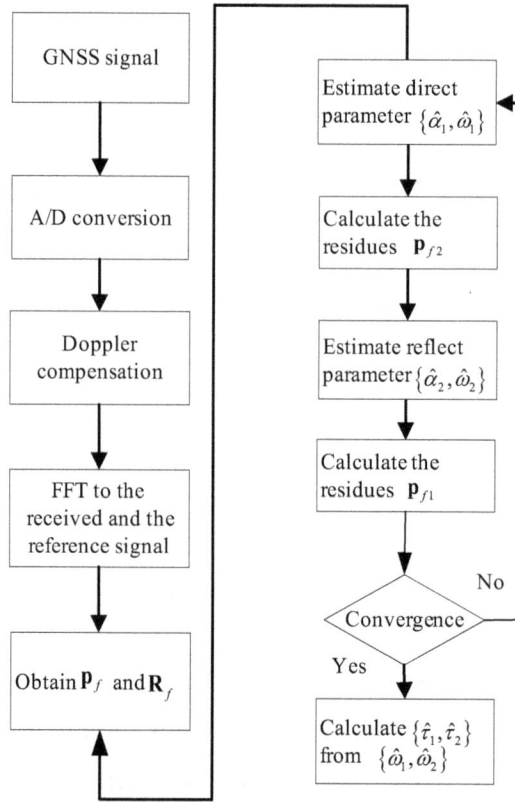

Figure 5. Diagram of the correlation domain code delay estimation algorithm

3.2.2. Code delay estimation in the data domain via NLS

Another DPE algorithm is proposed in this subsection. Different from the correlation domain algorithm, the unknown parameters $\{\alpha_p, \tau_p\}_{p=1}^2$ can also be estimated directly in the data domain, which can be described as follows.

Again, the signal after down-conversion is given by

$$y(n) = \sum_{p=1}^2 \alpha_p s(n - \tau_p) + e(n) \tag{51}$$

To obtain the unknown parameters $\{\alpha_p, \tau_p\}_{p=1}^2$ in equation (51), DFT is applied firstly

$$y_f(k) = s_f(k) \sum_{p=1}^2 \alpha_p e^{j\omega_p k} + e_f(k), -N/2 \le k \le N/2 - 1 \tag{52}$$

where $y_f(k)$, $s_f(k)$, $e_f(k)$ are the DFT of $y(n)$, $s(n)$ and $e(n)$ respectively. $\omega_p = -2\pi\tau_p f_s / N$ with f_s represents the sampling frequency. Till now, the code delay estimation is transformed to the angular frequency ω_p estimation, after which τ_p can be obtained by the relation $\tau_p = -\omega_p N / 2\pi f_s$.

To obtain $(\hat{\alpha}_p, \hat{\omega}_p)$, a NLS cost function as follows is defined

$$Q_1\left(\{\alpha_p, \omega_p\}^2_{p=1}\right) = \sum_{k=-N/2}^{N/2-1} \left| y_f(k) - s_f(k) \sum_{p=1}^{2} \alpha_p e^{j\omega_p k} \right|^2 \tag{53}$$

Let

$$\mathbf{S}_f = diag\left\{s_f(-N/2), s_f(-N/2+1), ..., s_f(N/2-1)\right\} \tag{54}$$

$$\mathbf{y}_f = \left[y_f(-N/2), y_f(-N/2+1), ... y_f(N/2-1)\right]^T \tag{55}$$

$$\mathbf{a}(\omega_p) = \left[e^{j\omega_p(-N/2)}, e^{j\omega_p(-N/2+1)}, ..., e^{j\omega_p(N/2-1)}\right]^T \tag{56}$$

Hence, minimizing the cost function in equation (53) is equivalent to minimizing the following cost function

$$Q_2\left(\{\alpha_p, \omega_p\}^2_{p=1}\right) = \left\| \mathbf{y}_f - \sum_{p=1}^{2} \alpha_p \mathbf{S}_f \mathbf{a}(\omega_p) \right\|^2 \tag{57}$$

Assume $\{\hat{\alpha}_q, \hat{\omega}_q\}^2_{q=1, q \neq p}$ is known prior or has been estimated, then we have

$$\mathbf{y}_{fp} = \mathbf{y}_f - \sum_{\substack{q=1 \\ q \neq p}}^{2} \hat{\alpha}_q \left[\mathbf{S}_f \mathbf{a}(\hat{\omega}_p)\right] \tag{58}$$

Substitute equation (58) into equation (57)

$$Q_2(\alpha_p, \omega_p) = \left\| \mathbf{y}_{fp} - \alpha_p \mathbf{S}_f \mathbf{a}(\omega_p) \right\|^2 \tag{59}$$

Equation (59) gets the minimum while $y_{fp} = \alpha_p S_f \mathbf{a}(\omega_p)$, then the estimation results $\hat{\omega}_p$ and $\hat{\alpha}_p$ can be obtained in the following way

$$\hat{\omega}_p = \arg\max_{\omega_p} \left| \mathbf{a}^H(\omega_p)\left(S_f^* y_{fp}\right)\right|^2 \tag{60}$$

$$\hat{\alpha}_p = \left. \frac{\mathbf{a}^H(\omega_p)\left(S_f^* y_{fp}\right)}{\left\|S_f\right\|_F^2} \right|_{\omega_p = \hat{\omega}_p} \tag{61}$$

From equation (60), $\hat{\omega}_p$ can be obtained as the location of the periodogram $\left|\mathbf{a}^H(\omega)\left(S_f^* y_{fp}\right)\right|^2$, which can be efficiently computed by using the FFT with the data sequence $S_f^* y_{fp}$ padded with zeros. (Note the padding with zeros is necessary to determine $\hat{\omega}_p$ with more accuracy.) Then $\hat{\alpha}_p$ is easily computed from the complex height of the peak of $\dfrac{\mathbf{a}^H(\omega_p)\left(S_f^* y_{fp}\right)}{\|S_f\|_F^2}$.

The diagram of the data domain code delay estimation algorithm is shown as in Figure 6.

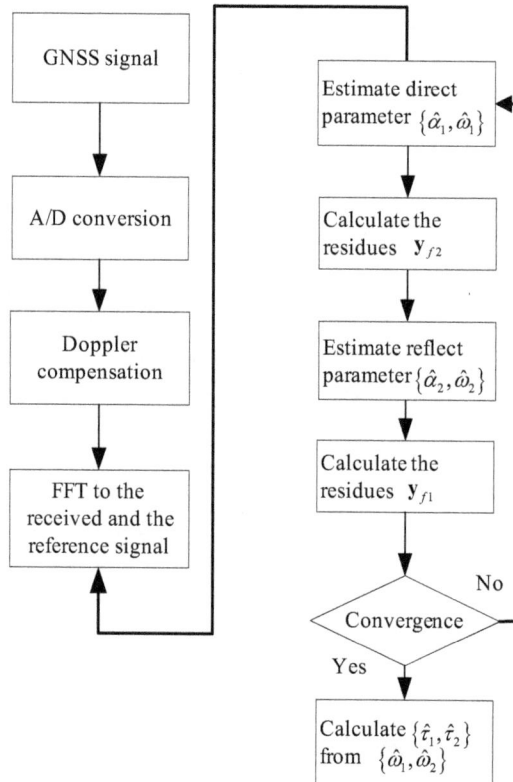

Figure 6. Diagram of the data domain code delay estimation algorithm

With the above simple preparations, we now proceed to present the relaxation algorithm for the minimization of the nonlinear least-squares cost function. The data domain WRELAX algorithm comprises the following steps.

Step 1. Assume $p=1$. Obtain $\{\hat{\omega}_1, \hat{\alpha}_1\}$ from equation (61) and equation (60).

Step 2. Assume $p=2$ (the LOS signal and one reflected path). Compute y_{f2} with (47) by using $\{\hat{\omega}_1, \hat{\alpha}_1\}$ obtained in Step (1). Obtain $\{\hat{\omega}_2, \hat{\alpha}_2\}$ from y_{f2} described above.

Next, compute y_{f1} with equation (47) by using $\{\hat{\omega}_2, \hat{\alpha}_2\}$ and redetermine $\{\hat{\omega}_1, \hat{\alpha}_1\}$ from y_{f1} as in equation (49) and equation (50) above.

Iterate the previous two substeps until "practical convergence" is achieved then we can obtain $\{\hat{\omega}_p, \hat{\alpha}_p\}_{p=1}^2$. Furthermore, by using $\hat{\omega}_p = -2\pi\hat{\tau}_p/NT_s$ the code delay $\hat{\tau}_p$, $p=1, 2$ can be obtained.

From the previous description one can find that the proposed algorithm can be implemented by simply FFT operation which deserves a less computation load.

3.2.3. Comparison of the above two algorithms

To further analyze the two proposed algorithms, the cost functions of them are further discussed in this subsection. Following the discussion in section 3.2.1 and section 3.2.2, we define

$$\mathbf{s}_f = \left[s_f(-N/2), s_f(-N/2+1), ..., s_f(N/2-1)\right]^T \tag{62}$$

Then

$$\mathbf{R}_f = \mathbf{s}_f \mathbf{s}_f^H \tag{63}$$

In the noise free situation we have

$$\mathbf{p}_{fp} = \alpha_p \mathbf{R}_f \mathbf{a}(\omega_p) \tag{64}$$

Then the correlation domain cost function in equation (49) can be further decomposed as

$$
\begin{aligned}
\hat{\omega}_p &= \arg\max_{\omega_p} \left| \mathbf{a}^H(\omega_p) \mathbf{R}_{fs}^* \mathbf{p}_{fp} \right|^2 \\
&= \arg\max_{\omega_p} \left| \mathbf{a}^H(\omega_p) \mathbf{R}_{fs}^* \left(\alpha_p \mathbf{R}_{fs} \mathbf{a}(\omega_p)\right) \right|^2 \\
&= \arg\max_{\omega_p} \left| \mathbf{a}^H(\omega_p) \left(\mathbf{s}_f \times \mathbf{s}_f^H\right)^* \left(\alpha_p \left(\mathbf{s}_f \times \mathbf{s}_f^H\right) \mathbf{a}(\omega_p)\right) \right|^2 \\
&= \arg\max_{\omega_p} \left| \mathbf{a}^H(\omega_p) \left[\left(\mathbf{s}_f^T \times \mathbf{s}_f^*\right) \times \left(\mathbf{s}_f \times \mathbf{s}_f^H\right)\right] \left(\alpha_p \mathbf{a}(\omega_p)\right) \right|^2
\end{aligned}
\tag{65}
$$

And in the same situation the data domain cost function in equation (60) can be deployed in the following way

$$\hat{\omega}_p = \arg\max_{\omega_p} \left| \mathbf{a}^H \left(\omega_p \right) \mathbf{S}_f^* \mathbf{y}_{fp} \right|^2$$

$$= \arg\max_{\omega_p} \left| \mathbf{a}^H \left(\omega_p \right) \mathbf{S}_f^* \left(\alpha_p \mathbf{S}_f \mathbf{a} \left(\omega_p \right) \right) \right|^2 \tag{66}$$

$$= \arg\max_{\omega_p} \left| \mathbf{a}^H \left(\omega_p \right) \left(\mathbf{s}_f^* \mathbf{s}_f^T \right) \left(\alpha_p \mathbf{a} \left(\omega_p \right) \right) \right|^2$$

It has been discussed in previous subsection that $\hat{\omega}_p$ can be obtained by using the FFT with the data sequence $\mathbf{R}_f \mathbf{p}_{fp}$ and $\mathbf{S}_f{}^H \mathbf{y}_{fp}$, respectively. Note $(\alpha_p \mathbf{a}(\omega_p))$ is an impulse signal, and $\text{FFT}(\mathbf{s}_f^* \mathbf{s}_f^T)$ convolutions with the impulse signal is still $\text{FFT}(\mathbf{s}_f^* \mathbf{s}_f^T)$ but with a code delay, which is the parameter to be estimated. From equation (65) and (66) we realized that the cost function are the FFT of $\alpha_p \mathbf{a}(\omega_p)$ weighted by different window function, where window function in the correlation domain is $w_1 = \text{FFT}((\mathbf{s}_f^T \mathbf{s}_f^*)(\mathbf{s}_f \mathbf{s}_f^H))$, but the window function in data domain is $w_2 = \text{FFT}(\mathbf{s}_f^* \mathbf{s}_f^T)$. It's clear that the weighted window w_1, w_2 will broaden the cost function, and the impact of w_1 is more serious than that of w_2, which can also be seen from the comparison of the cost function given in Figure 7. From this point of view, the proposed data domain algorithm would deserve a more accurate estimation result.

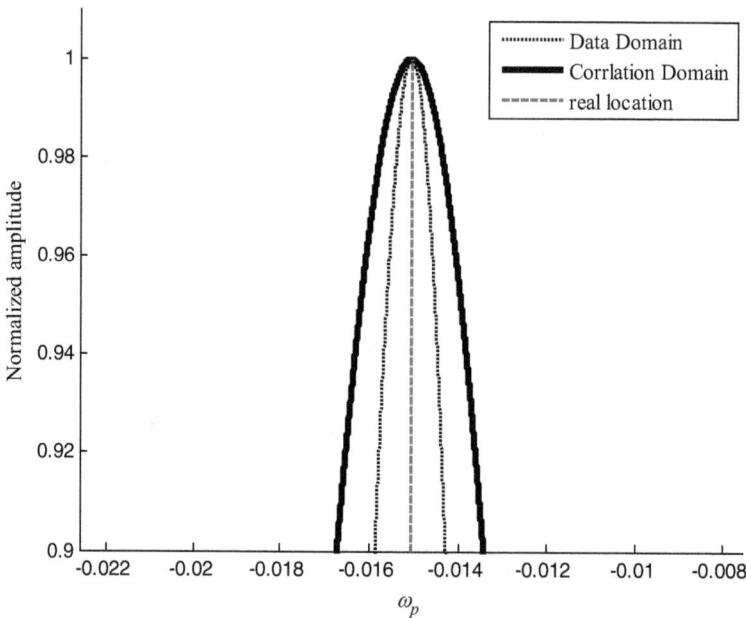

Figure 7. Comparison of the correlation domain and the data domain cost function

Correlation of the received data and the locally generated signal and the auto-correlation of the locally generated signal should be calculated in correlation domain WRELAX algorithm.

Suppose the iteration times of the two proposed algorithms are the same, the correlation domain WRELAX algorithm is more complex than the data domain WRELAX algorithm.

3.2.4. Numerical results

To investigate the performance of the proposed algorithm in the presence of multipath, a simulation experiment was performed. The code delay estimation error can be represented as the function of τ, and as explained in [l] and [7], maximum and minimum errors occur when the multipath signal is in-phase $\Delta\varphi = 0°$ or out-of-phase $\Delta\varphi = \pm 180°$ with the LOS signal. The curve of the maximum and minimum errors is regarded as the error envelopes, which is used in the following numerical experiment to evaluate the performance.

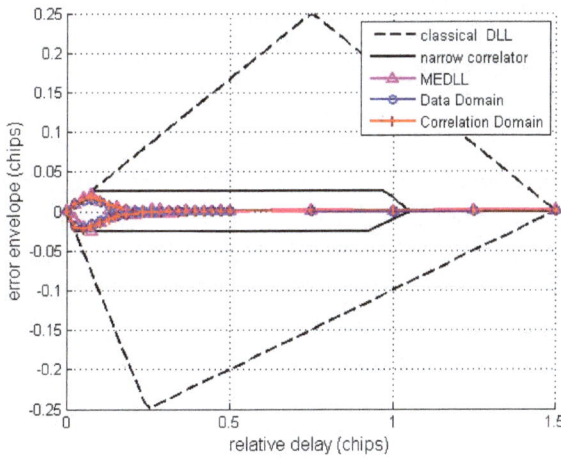

(a). error envelopes

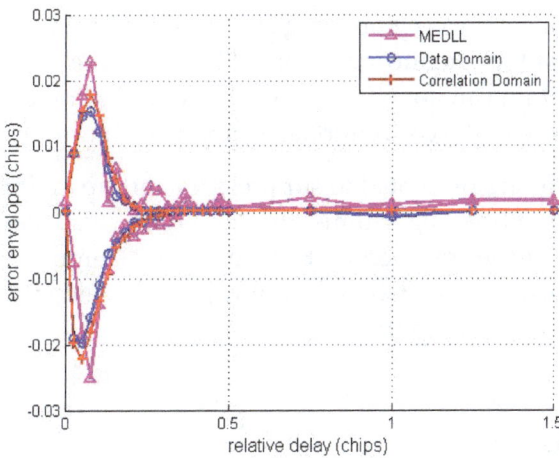

(b). details of (a)

Figure 8. Comparison of the error envelopes

Suppose there is only one reflected multipath signal combined with the LOS signal. The noise and the receiver bandwidth are not considered in the experiment. The relative amplitude of the direct and the reflected signal is $a_2/a_1=0.5$. The code delay difference $\Delta\tau=\tau_2-\tau_1$ is varied from 0 to 1.5 chips. The error is calculated at the maximum points when the multipath signal is at 0°in phase with respect to the LOS signal. The results are shown in Figure 8(a).

In Figure 8, the curve 'classical DLL' denotes the error envelope of the classical DLL, and the curve 'narrow correlator' denotes the error envelope of the narrow correlator with spacing $d=0.1$ chip. Both the 'classical DLL' and the 'narrow correlator' were calculated by taking the DLL equations and solving for the tracking error. The curve 'MEDLL', 'Correlation Domain' and 'Data Domain' denote the code delay estimation error of MEDLL, proposed correlation domain algorithm and data domain algorithm, respectively. The results indicate that in the presence of multipath, the classical DLL technique has a large tracking error, which fails in multipath mitigation. The tracking error is reduced in the narrow correlator. However, based on the previous conclusion, the tracking error is nearly constant when the correlation spacing is less than 0.1 chip. The tracking accuracy of the algorithms in this section and MEDLL show a considerable improvement over the narrow correlator technique and the classical DLL technique. The improved performance can be of great help in critical GPS applications where the multipath errors associated with using conventional receivers can easily exceed the accuracy requirements. To further compare the performance of the two proposed algorithms and MEDLL, we magnify the results in Figure 8(a) to obtain Figure 8(b). It is clear that the proposed two algorithms have a more favorable performance than MEDLL. Furthermore, the proposed data domain algorithm is superior to the correlation domain algorithm, which is consistent with the conclusion in section 3.2.3.

4. Conclusion

A novel acquisition algorithm is firstly proposed in this chapter. In the proposed algorithm, the Doppler frequency is obtained by utilizing FFT of the squared data. Then, the PRN index and initial code phases of the satellite are obtained based on the NLS criterion. It can be seen that the proposed algorithm can not only reduce the computation complexity but also attain a comparable performance to the conventional acquisition algorithm.

After that, two algorithms are presented to suppress the multipath interference by estimating the code delay. In the proposed algorithms, the code delay was obtained by solving a NLS equation, which can be further realized by FFT operation. Compared with the conventional estimation algorithm, the proposed two algorithms perform better in multipath propagating environments and bear lower computation burden.

Acknowledgements

The work of this chapter is supported by the Project of the National Natural Science Foundation of China (Grant No. 61179064, 61172112, 61271404 and 61471363), and the Fundamental Research Funds for the Central Universities (Grant 3122014D008 and 3122014B001).

Author details

Renbiao Wu[1], Qiongqiong Jia[1], Wenyi Wang[1] and Jie Li[2]

1 Tianjin Key Lab for Advanced Signal Processing, Civil Aviation University of China, Tianjin, P. R. China

2 Department of Electronic and Information Engineering, Zhonghuan Information College Tianjin University of Technology, Tianjin, P. R. China

References

[1] Kalyanaraman S K, Braasch M S, Kelly J M. Code Tracking Architecture Influence on GPS Carrier Multipath. IEEE Transactions on Aerospace and Electronic Systems 2006; 42(2) 548-561.

[2] Kos T, Markezic I, Pokrajcic J. Effects of Multipath Reception on GPS Positioning Performance. 52nd International Symposium ELMAR-2010: conference proceedings, 15-17 September 2010, Zadar, Croatia. 399-402.

[3] Scire-Scappuzzo F, Makarov S N. A Low-Multipath Wideband GPS Antenna With Cutoff or Non-Cutoff Corrugated Ground Plane. Antennas and Propagation 2009; 57(1): 33-46

[4] Maqsood M, Gao S, Brown T, et al. A Compact Multipath Mitigating Ground Plane for Multiband GNSS Antennas. IEEE TRANSACTIONS ON ANTENNAS AND PROPAGATION 2014; 61(5) 2775-2782.

[5] Ray J K, Cannon M E, Fenton P. GPS Code and Carrier Multipath Mitigation Using a Multiantenna System. IEEE Transactions on Aerospace and Electronic Systems 2001; 37(1) 183-195.

[6] Alfred R L. GPS Landing System Reference Antenna. IEEE Antennas and Propagation Society 2010; 52(1) 104-113.

[7] Michael S B. GPS Multipath Model Validation. IEEE Position Location and Navigation Symposium 1996: conference proceedings, April 1996, Atlanta, GA. 672-678.

[8] Cannon M E, Lachapelle G, Qiu W, et al. Performance Analysis of a Narrow Correlator Spacing Receiver for Precise Static GPS Positioning. IEEE Position Location and Navigation Symposium 1994: conference proceedings, April 1994, Las Vegas. 2994—2999

[9] Tung Hai Ta, Son Hong Ngo. A Novel Signal Acquisition Method for GPS Dual-Frequency Ll C/A and L2C Receivers. International Conference on Advanced Technologies for Communications (ATC 2011) : conference proceedings, Aug 2011. 231-234.

[10] Chengjun Li, Mingquan Lu, Zhenming Feng, Qi Zhang. Study on GPS Acquisition Algorithm and Performance Analysis. Journal of Electronics & Information Technology 2010; 32(2) 296-300.

[11] Olivier Besson, Peter Stoica. Nonlinear least-squares frequency estimation and detection for sinusoidal signals with arbitrary envelope. Digital Signal Processing 1999; 9(1) 45-56.

[12] Jian Li, Renbiao Wu. An efficient algorithm for time delay estimation. IEEE Transactions on Signal Processing 1998; 46(8) 2231-2235.

[13] Jian Li, Dunmin Zheng, Peter Stoica. Angle and Waveform Estimation via RELAX. IEEE Transactions on Aerospace and Electronic Systems 1997; 33(3) 1077-1087.

[14] Renbiao Wu, Jian Li, Zhengshe Liu. Super resolution time delay estimation via MODE-WRELAX. IEEE Transactions on Aerospace and Electronic Systems 1999; 35(1) 294-307.

[15] Renbiao Wu, Jian Li. Time-Delay Estimation via Optimizing Highly Oscillatory Cost Functions. IEEE Journal of Oceanic Engineering 1998; 23(3) 235-244.

6

Use of Fast Fourier Transform for Sensitivity Analysis

Andrej Prošek and Matjaž Leskovar

Additional information is available at the end of the chapter

1. Introduction

The uncertainty quantification of code calculations is typically accompanied by a sensitivity analysis, in which the influence of the individual contributors to the uncertainty is determined. In the sensitivity analysis, the basic step is to perform sensitivity calculations varying the input parameters. One or more input parameters could be varied at a time. The typical statistical methods for the sensitivity analysis used in uncertainty methods are for example Pearson's Correlation Coefficient, Standardized Regression Coefficient, Partial Correlation Coefficient and others [1]. The output results are time domain signals. The objective of this study was to use fast Fourier transform (FFT) based approaches to determine the sensitivity of output parameters. In the reference [2], the FFT based approaches have been used for the accuracy quantification. The difference between the accuracy quantification and the sensitivity analysis is that in the accuracy quantification the experimental data are compared to the code calculated data, while in the sensitivity analysis the reference calculation signal is compared to the sensitivity run calculation signal. To do this comparison, first the fast Fourier transform is used to transform time domain signals into frequency domain signals. Then, the average amplitude is calculated, which is the sum of the amplitudes of the frequency domain difference signal (between the sensitivity run calculation signal and the reference calculation signal) normalized by the sum of the amplitudes of the frequency domain reference signal. Finally, the figures of merit based on the average amplitude are used to judge the sensitivity.

Such a FFT based approach is different from the typical sensitivity analysis using a statistical procedure to determine the influence of sensitive input parameters on the output parameter. Namely, the influence of the sensitive input parameters is represented by the average amplitude which remembers the previous history.

In this study it is shown that the proposed FFT based tool is complementary and a good alternative to the mentioned typical statistical methods, if one parameter is varied at a time.

The advantage of the proposed method is that the results of the sensitivity analysis obtained by the FFT based tool could be ranked. It provides a consistent ranking of sensitive input parameters according to their influence to the output parameter when one parameter is varied at a time, and based on the fact that the same method can be used for more participants performing calculations. For example, there is no need to have ranking levels like it was proposed in the Phase III of the Best Estimate Methods – Uncertainty and Sensitivity Evaluation (BEMUSE) programme [3] for the qualitative judgement, because in this approach the figure of merit is a quantitative value and therefore can be directly ranked. The zero value of the figure of merit means a not relevant sensitive parameter and the larger the figure of merit is the more influential the input parameter is.

The difference between statistical methods and the FFT based approach is that in the case of statistical methods the influence of each varied input parameter on the output result can be obtained even if more parameters are varied at a time. This cannot be done by FFT based tools when more parameters are varied at a time. Rather, the total influence of sensitive parameters on the result is given by the FFT based tool. But the good thing is that the same measure is used for both single and multiple variations and in this way the compensation effects of the influence of different sensitive parameters could be studied. As was already mentioned the FFT based approaches are complementary to statistical methods.

In this Chapter, first the original fast Fourier transform based method (FFTBM) approach is described. The average amplitude, the signal mirroring, the FFTBM Add-in tool and the time dependent accuracy measures are introduced. Based on signal mirroring the improved FFTBM by signal mirroring (FFTBM-SM) was developed. By calculating the time dependent average amplitude it can be answered, which discrepancy due to the parameter variation contributes to the sensitivity and how much is its contribution. The past application of the FFT based tool for the accuracy quantification showed that the original FFTBM gave an unrealistic judgment of the average amplitude for monotonically increasing or decreasing functions, causing problems in the FFTBM results interpretation. It was found out [4] that the reason for such an unrealistic calculated accuracy of increasing/decreasing signals is the edge between the first and last data point of the investigated signal, when the signal is periodically extended. Namely, if the values of the first and last data point of the investigated signal differ, then there are discontinuities present in the periodically extended signal seen by the discrete Fourier transform, which views the finite domain signal as an infinite periodic signal. The discontinuities give several harmonic components in the frequency domain, thus increasing the sum of the amplitudes, on which FFTBM is based, and by this influencing the accuracy. The influence of the edge due to the periodically extended signal is for clarity reasons called the edge effect.

Then the methods used for the sensitivity analysis are described. For the demonstration of the sensitivity study using FFT based tools the L2-5 test, which simulates the large break loss of coolant accident in the Loss of Fluid test (LOFT) facility, was used. The signals used were obtained from the Organisation for Economic Co-operation and Development (OECD) BEMUSE project. In the BEMUSE project there were 14 participants, each performing a reference calculation and 15 sensitivity runs of the LOFT L2-5 test. Three output parameters were provided: the upper plenum pressure, the primary side mass inventory and the rod surface temperature.

Finally, the application of the FFT based approaches to the sensitivity analysis is described. Both FFTBM and FFTBM-SM were used.

2. Fast fourier transform based method description

The FFT based method is proposed for the sensitivity analysis, which is analogous to the FFT based code-accuracy assessment described in ref. [5]. The FFT based approach for code accuracy consists of three steps: a) selection of the test case (experimental or plant measured time trends to compare), b) qualitative analysis, and c) quantitative analysis. The qualitative analysis is necessary before quantifying the discrepancies between the measured and calculated trends. The qualitative analysis includes also the visual observation of plots and the evaluation of the discrepancies between the measured and calculated trends, which should be predictable and understood. For the sensitivity analysis the same FFT based approach is used for the quantitative analysis as used for the code-accuracy. However, the signals compared now are the output signal obtained with the reference value of the input parameter and the output signal as the result of the sensitive input parameter variation.

In the quantitative analysis, the influence of the sensitive input parameter variation is judged in the frequency domain. Therefore the time domain signals used in the sensitivity analysis have to be transformed in the frequency domain signals. The addressed time domain signals assume values different from zero only in the interval [0, T_d], where T_d is the duration of the signal. Also the digital computers can only work with information that is discrete and finite in length (e.g. N points) and there is no version of the Fourier transform that uses finite length signals [6]. The way around this is to make the finite data look like an infinite length signal. This is done by imagining that the signal has an infinite number of samples on the left and right of the actual points. The imagined samples can be a duplication of the actual data points. In this case, the signal looks discrete and periodic. This calls for the discrete Fourier transform (DFT) to be used. There are several ways to calculate DFT. One method is FFT. While it produces the same results as the other approaches, it is incredibly more efficient. The key point to understand the FFTBM is that the periodicity is invoked in order to be able to use a mathematical tool, i.e., the DFT. It seems that the developers of the original FFTBM have not been sufficiently aware of this fact.

The discrete Fourier transform views both, the time domain and the frequency domain, as periodic [6]. However, the signals to be used for the comparison are not periodic and the user must conform to the DFT's view of the world. When a new period starts, the N samples on the left side are not related to the samples on the right side. However, DFT views these N points to be a single period of an infinitely long periodic signal. This means that the left side of the signal is connected to the right side of the signal, and vice versa. The most serious consequence of the time domain periodicity is the occurrence of the edge, where the signals are glued. When the signal spectrum is calculated with DFT, the edge is taken into account, despite the fact that the edge has no physical meaning for the comparison, since it was introduced artificially by the applied numerical method. It is known that the edge produces a variegated spectrum of frequencies due to the discontinuity of the edge. These frequencies originating from the artificially introduced edge may overshadow the frequency spectrum of the investigated signal. Therefore an improved version of FFTBM by signal mirroring has been proposed, which is described in detail in [4]. Both the original FFTBM and the improved FFTBM by signal mirroring have been used in the demonstration application. The same equations are used for the calculation of the average amplitude, like for the original FFTBM, except that, instead of the original signals, the symmetrized signals are used (for further details see ref. [4]). In the following it is first described how the average amplitude is calculated.

2.1. Average amplitude

FFT is another method for calculating the DFT. While it produces the same result as the other approaches, it significantly reduces the computation time. FFT usually operates with a number of values N that is a power of two. Typically, N is selected between 32 and 4096 [6]. In addition, the sampling theorem must be fulfilled to avoid the distortion of sampled signals due to the aliasing occurrence. The sampling theorem says: "a signal that varies continuously with time is completely determined by its values at an infinite sequence of equally spaced times if the frequency of these sampling times is greater than twice the highest frequency component of the signal" [7]. Thus if the number of points defining the function in the time domain is $N=2^{m+1}$, then according to the sampling theorem the sampling frequency is:

$$\frac{1}{\tau} = f_s = 2f_{max} = \frac{N}{t_d} = \frac{2^{m+1}}{t_d} \tag{1}$$

where τ is the sampling interval, t_d is the transient time duration of the sampled signal and f_{max} is the highest (maximum) frequency component of the signal. The sampling theorem does not hold beyond f_{max}. From the relation in Eq. (1) it is seen that the number of points selection is strictly connected to the sampling frequency. The FFT algorithm requires the number of points, equally spaced, which is a power with base 2. Generally an interpolation is necessary to satisfy this requirement. The original FFTBM is done so that the default value of the exponent m ranges from 8 to 11. This gives N ranging from 512 to 4096. The final number of points used by FFTBM is determined depending on the value of f_{fix}, which is the minimum requested maximum frequency and is input value. If f_{max} is not larger than f_{fix}, the number of points is doubled (exponent m is increased for 1) until the criterion is satisfied or the exponent m equals to 11. Please note, that the minimum value of the exponent m is 8. The FFTBM application implies the following input values: the fixed frequency f_{fix} (minimum maximum frequency of the analysis, this determines the number of points N), the cut off frequency (f_{cut}), the start time t_s and the end time t_e of the analysed window (determines the analysis window $t_d = t_s - t_e$). A cut off frequency has been introduced to cut off spurious contributions, generally negligible. When f_{cut} is equal or larger than f_{fix}, all frequency components are considered.

For the calculation of the differences between the output signal obtained with the reference value of the input parameter (reference signal $F_{ref}(t)$) and the output signal as the result of the sensitive input parameter variation (sensitive signal $F_{sen}(t)$), the reference signal ($F_{ref}(t)$) and the difference signal $\Delta F(t)$ are needed. The difference signal in the time domain is defined as:

$$\Delta F(t) = F_{ref}(t) - F_{sen}(t). \tag{2}$$

After performing the fast Fourier transform the obtained spectra of amplitudes are used for the calculation of the average amplitude (AA):

$$AA = \frac{\sum_{n=0}^{2^m}|\tilde{\Delta}F(f_n)|}{\sum_{n=0}^{2^m}|\tilde{F}_{ref}(f_n)|}, \tag{3}$$

where $|\tilde{\Delta}F(f_n)|$ is the difference signal amplitude at frequency f_n and $|\tilde{F}_{ref}(f_n)|$ is the reference signal amplitude at frequency f_n. The AA factor can be considered as a sort of average fractional difference and the closer the AA value is to zero, the smaller is the sensitivity (influence). In our specific application, the larger the sensitivity is the larger is the difference between the signals, normally resulting in a larger AA value. Typically, based on the previous experience in the accuracy quantification the values of AA below 0.3 indicate a small influence (in the case of pressure below 0.1), while the values above 0.5 indicate a large influence.

The above Eq. (3) can be also viewed as:

$$AA = \frac{AA_{dif}}{AA_{ref}}, \tag{4}$$

where AA_{dif} is the average amplitude of the difference signal and AA_{ref} is the average amplitude of the reference signal:

$$AA_{dif} = \frac{1}{2^m+1}\sum_{n=0}^{2^m}|\tilde{\Delta}F(f_n)|, \tag{5}$$

$$AA_{ref} = \frac{1}{2^m+1}\sum_{n=0}^{2^m}|\tilde{F}_{ref}(f_n)|. \tag{6}$$

If the reference and sensitive signals are the same, the difference signal is zero. The larger the difference is, the larger is AA_{dif} (in principle). On the other hand, AA_{ref} normalizes the average amplitude AA and the higher the sum of amplitudes is, the lower is the average amplitude AA. This means that for ranking the sensitivities of the selected output parameter only AA_{dif} has an influence. On the other hand, for judging the influence of a single sensitive parameter variation on different output parameters, besides AA_{dif} also AA_{ref} influences the ranking due to the different AA_{ref} the output parameters have.

$$fraction \; A0 = \frac{\left|\tilde{\Delta}F(f_0)\right|}{\sum\limits_{n=0}^{2^m}\left|\tilde{\Delta}F(f_n)\right|}, \tag{7}$$

where $\left|\tilde{\Delta}F(f_0)\right|$ is the mean value of the difference signal and it is equivalent to the mean error (ME) in the time domain as defined in ref. [8]. The measure *fraction A0* shows when the frequency amplitudes are dominating or when the mean values (like constant differences) are dominating the sum of the amplitudes.

2.2. Signal mirroring

If we have a function $F(t)$ where $0 \leq t \leq t_d$ and t_d is the transient time duration, its mirrored function is defined as $F_{mir}(t) = F(-t)$, where $-t_d \leq t \leq 0$. From these functions a new function is composed which is symmetrical in regard to the y-axis: $F_m(t)$, where $-t_d \leq t \leq t_d$. This is illustrated in Figure 1. By composing the original signal (shown in Figure 1(a)) and its mirrored signal (signal mirroring), a signal without the edge between the first and the last data sample is obtained, and it is called symmetrized signal (shown in Figure 1(b)). It has the double number of points in order not to lose any information. Also it should be noted that the edge is not present in the original time domain signal (see Figure 1(a)). However, when performing FFT, the aperiodic original signal is treated as a periodic original signal as mentioned before and therefore the edge is part of the periodic original signal, what is not physical. In the case of the symmetrized signal the edge is not present even when treating the signal as periodic.

For the calculation of the average amplitude by signal mirroring AA_m the Eq. (3) is used like for the calculation of AA, except that, instead of the original signals, the symmetrized signals are used. This may be efficiently done by signal mirroring, where the investigated signal is mirrored before the original FFTBM is applied. By composing the original signal and its mirrored signal (signal mirroring), a symmetric signal (also called symmetrized signal) with the same characteristics is obtained, but without introducing the edge when viewed as an infinite periodic signal (for details refer to refs. [4, 9]). FFTBM using the symmetrized signals is called FFTBM-SM.

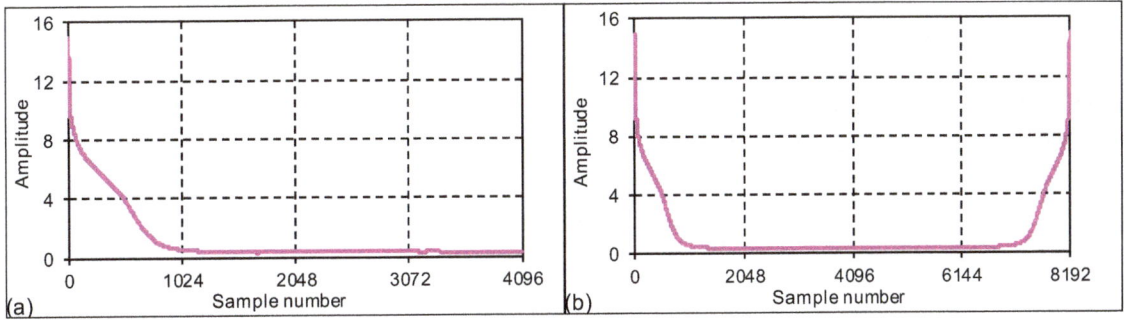

Figure 1. Examples of (a) original digital signal and (b) symmetrized digital signal

2.3. FFTBM Add-in tool

To take the advantages of spread sheets in preparing input forms, analysing data (including analysis of values), modifying graphs and the capability to store time recorded data, plots, input forms and results, the Jožef Stefan Institute (JSI) in-house Microsoft Excel Add-in for the accuracy evaluation of thermal-hydraulic code calculations with FFTBM has been developed in 2003 [10]. Later the tool was upgraded with the capability to symmetrize the signals, and some other improvements. The upgraded tool is called JSI FFTBM Add-in 2007 [11] and it has been used for the sensitivity study, described in this chapter. It includes both FFTBM and FFTBM-SM. As already mentioned, the difference between FFTBM and FFTBM-SM is that in the latter the signals are symmetrized to eliminate the edge effect in calculating the average amplitude by signal mirroring (AA_m).

JSI FFTBM Add-in 2007 provides additional information on interpolated data of the signals used, the difference signals, the amplitude spectra and the AA dependency on the cut frequency. The user can use the interpolated data for visual checking about the agreement of the original signal and the interpolated signal. The amplitude spectra give the possibility to compare the spectra between different signals. Information on the AA dependency on the cut frequency is used to check if the cut frequency is selected properly. Usually the dependency is not so big, therefore by default AA at the selected f_{cut} frequencies is calculated:

- minimal AA (AA_{min}) at frequency (when $f_{cut} > 0.05 f_{max}$) which gives the minimum AA,
- average AA (AA_{avg}) is calculated as the average AA at all cut frequencies,
- maximal AA (AA_{max}) at the frequency (when $f_{cut} > 0.05 f_{max}$) which gives the maximum AA,
- 5 percentile AA (AA_{05}) at frequency $f_{cut} = 0.05 f_{max}$,
- 50 percentile AA (AA_{50}) at frequency $f_{cut} = 0.5 f_{max}$,
- 100 percentile AA (AA_{100}) considering all amplitudes (for $f_{cut} = f_{max}$).

This gives an indication if AA is significantly dependent on the cut frequency. In principle, AA should not be much different when half or all frequencies from the amplitude spectrum are considered for the AA calculation. If this is not the case, deeper insight into AA is needed to judge if there is a spurious contribution present in the signal. For example, in the case of measured data, noise may be present in the signal. In the case of code calculated digital data, noise is not present in the signals, therefore the whole amplitude spectrum is recommended for the AA calculation. Nevertheless, the user should be aware that AA depends on the cut frequency and that the result may change when not all frequencies are considered. Typically higher frequency components have lower amplitudes than lower frequencies, therefore the lower frequency content is always used for the AA calculation (only higher frequency components are filtered).

2.4. Time dependent average amplitude

In the ref. [12] the influence of the time window selection was studied. Instead of a few phenomenological windows a series of narrow windows (phases) could be selected. This gives the possibility to get the time dependency of the average amplitude. The increasing time interval was defined as a set of time intervals each increased for the duration of one narrow time window and the last time interval being the whole transient duration time. By increasing the time interval we see how the average amplitude changes with the time progression as it is shown in Figure 2. The average amplitude was calculated by the original FFTBM not considering the edge effect. Therefore the average amplitude shown in Figure 2(b) first increases and then partly decreases in spite of the discrepancy present all this time during the temperature increase shown in Figure 2(a).

The time dependent average amplitude is also indispensable for the sensitivity analysis. From such a time dependant average amplitude it can be easily seen when the largest influence occurs on the output parameter due to the sensitive parameter variation.

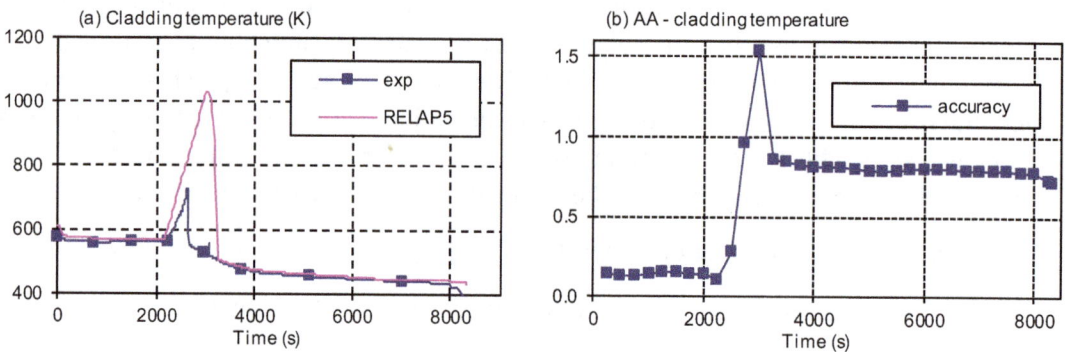

Figure 2. Time trend of (a) primary pressure and (b) the corresponding average amplitude

3. Methods used for sensitivity analysis

In BEMUSE, Phase II, single parameter sensitivity analyses have been proposed and performed by the participants to study the influence of different parameters (break area, gap conductivity, core pressure drops, time of scram etc.) upon the predicted large-break loss-of-coolant accident (LBLOCA) evolution [13]. The performed sensitivity studies were intended to be used as guidance for deriving uncertainties of relevant input parameters like for phase III of the BEMUSE programme.

The sensitivity analysis is concerned, generally, with the influence of inputs on the output and output variability. Generally, sensitivity analyses are conducted by defining the model and its independent and dependent variables, assigning probability density functions to each input parameter, generating an input matrix through an appropriate random sampling method, performing calculations, and assessing the influences and relative importance of each input/ output relationship.

In our demonstration case, the calculated data obtained in the Phase II of BEMUSE provided by the host organization have been used. The input matrix consists of the single parameter variation. The range of variations has been proposed by the host organization. For sensitivity calculations, only one value (minimal or maximal) of the input parameter was proposed when the range of the parameter variation was specified for selected sensitive parameter. For each single parameter variation the calculation was performed. The influence of the sensitive parameter variation has been estimated through the application of FFT based approaches.

We will call sensitivity (of the output parameter Y versus the i-th input parameter X_i) a measure having the dimension of $\frac{\partial Y}{\partial X_i}$ that is independent of the range of variation of the parameter X_i. Sensitivity is related to output variability.

We will call influence a measure of the effect of the variation of the parameter X_i on its full range (ΔX_i) having the dimension of $\frac{\partial Y}{\partial X_i} \Delta X_i$ (same as Y) or more often dimensionless form $\frac{\partial Y}{\partial X_i} \frac{\Delta Y}{\Delta X_i}$.

In our sensitivity analysis the influences were determined. Please note that classical measures of influences are: Pearson's or Spearman's Correlation Coefficients, Standardised Regression Coefficients, etc. [14]. In FFT based approaches the AA is a dimensionless number, showing influences in terms of the average amplitude obtained in the frequency domain which represents the physical influence (e.g. temperature or pressure change).

3.1. Test description

The LOFT L2-5 test was selected for this demonstration because a huge amount of data was available [15]. The reference and sensitivity calculations of the LOFT L2-5 test were performed in the phase II of the BEMUSE research program. The nuclear LOFT integral test facility is a scale model of a pressurized water reactor. The objective of the test was to simulate a loss of coolant accident (LOCA) caused by a double-ended, off-shear guillotine cold leg rupture coupled with a loss of off-site power. The experiment was initiated by opening the quick opening blowdown valves in the broken loop hot and cold legs. The reactor scrammed and emergency core cooling systems started their injection. After initial heatup the core was quenched at 65 s, following the core reflood. The low pressure injection system (LPIS) was stopped at 107.1 s, after the experiment was considered complete. In total 14 calculations from 13 organizations were performed. For more detailed information on the calculations the reader is referred to [4, 15].

3.2. Sensitivity calculations description

The series of sensitivity calculations with assigned parameters was proposed to participants. For each parameter the host organization recommended the value to be used. The short description of the cases to be analysed is given in Table 1.

ID	Parameter	Recommended values (RV)	Description
S-1	Break Area	RC x 1.15	Tube diameter from reactor pressure vessel to break point shall be modified in respect to RC.
S-2	Gap Conductivity	RC x 0.2	Only in the hot rod in the hot channel (zone 4).
S-3	Gap Thickness	RC x 3	Only in the hot rod in the hot channel (zone 4).
S-4	Presence of Crud	0.15 mm	Consideration of 0.15 mm of crud in hot rod in hot channel with thermal conductivity that is characteristic of ceramic material, e.g. Al2O3.
S-5	Fuel Conductivity	RC x 0.4	Only in the hot rod in the hot channel (zone 4).
S-6	Core Pressure Drop	RC + D: $DP_{tot}=(DP_{tot})_{RC} \times 1.3$	The pressure drop across the core shall increase (decrease) of an amount D to obtain a total pressure drop that is 30% bigger than the total pressure drop of the reference case.
S-7	CCFL at Upper Tie Plate (UTP) and/or connection upper plenum (UP)	Range not assigned	Counter courant flow limitation (CCFL) is nodalization dependent. Each participant can propose a solution.
S-8	Decay Power	RC x 1.25	The decay power has to be 25% bigger than in the reference case.
S-9	Time of Scram	RC + 1 s	The power curve shall follow the imposed trend that implies full power till RC + 1 sec and after that shall followed the decay power.
S-10	Maximum Linear Power	RC x 1.5	Only in the hot rod in the hot channel (zone 4).
S-11	Accumulator Pressure	RC - 0.5 MPa	Set point of accumulator pressure 0.5 MPa lower than the set point in the base case (= 4.29 MPa).
S-12	Accumulator Liquid Mass	RC x 0.7	Accumulator liquid mass shall be 0.7 times the value in the reference case.
S-13	Pressurizer Level	RC - 0.5 m	Pressurizer level shall be 0.5 m lower than in the reference case.
S-14	HPIS	Failure	Failure of HPIS.
S-15	LPIS injection initiated	RC + 30 s	Delay in starting LPIS injection.

RC: value used in Reference Case

Table 1. List of sensitivity analyses and proposed parameter variations (adapted per Table 6 in ref. [15])

In Table 2 sensitivity calculations performed by 14 participants are shown. Each row presents sensitivity calculations S-1 to S-15. If the calculation is performed with recommended values, the sign √ is used. If another value has been used, the value of the sensitive parameter is indicated. If the sensitivity calculation has not been performed, the cell is shaded grey.

Name	S-1	S-2	S-3	S-4	S-5	S-6	S-7	S-8	S-9	S-10	S-11	S-12	S-13	S-14	S-15
P-1	RC × 1.15 (area)	RC × 0.5	√	√	RC × 0.5	√	CCFL at UTP	√	√	√	√	√	√	√	√
P-2	√*	√*	√*		√*	√*		√*	√*	√*	√*	√	√*	√*	√*
P-3	RC × 0.7	RC × 2	√*	√*	√*	√*		√*	√*	√*	√*	√*	√*	√*	√*
P-4	√		√		√	√		√	√	√	√	√	√	√	√
P-5	Adjusted to pressure	√*				RC × 2 (resistance)		√	√	√					
P-6	RC × 1.15 (open size)	√	√	√*	√*	√*	√*	√*	√*	√*	√*	√*	√*	√*	√*
P-7	√*					√*		√*		√*	√*				
P-8	√	RC × 0.4	RC × 2.4	√	RC × 0.6	√	√	√	√	RC × 1.2	√	√	√	√	√
P-9	√*	√*	√*	√*	√*	√*	CCFL at UP	√*	√*	√*	√*	√*	√*	√*	√*
P-10	√	√	√		√	√		√	√	√	√	√	√	√	√
P-11	√*	√*	√*	√*	√*	√*		√*	√*	√*	√*	√*	√*	√*	√*
P-12	√	√	√	√	√	√	turned off	√	√	√	√	√	√	√	√
P-13	RC × 1.15 (area)	√	√	√	√	√	default	√	√	√	√	√	√	√	√
P-14	√	√	√	√	√	√	√	√	√	√	√	√	√	√	√

* no information given in Ref. [15], but high certainty that recommended values were used;

√ - recommended value used

Table 2. Sensitivity calculations performed by participants

3.3. Figures of merit for sensitivity analysis

Same figures of merit were proposed for the original FFTBM and improved FFTBM-SM. In the case of signal mirroring the additional index m is used to distinguish the FFTBM-SM from FFTBM. The first figure of merit is the average amplitude (AA or AA_m), which tells how the single input parameter variation (or combination of input parameter variations) influences the output parameter. As there are several sensitive input parameters, several participants performing sensitivity runs and more selected output parameters, three additional figures of merit were proposed. The average amplitude of the participant sensitivity runs (AA_p or AA_{mp}) is used to judge which sensitivity runs set is more influential to the input parameter variations. AA_p or AA_{mp} is calculated as the average of AAs for the participant sensitivity runs (15 in our specific case). The average amplitude of the sensitivity runs for the same sensitive parameter (AA_s or AA_{ms}) is used to judge how influential (in average) the sensitive parameter is in calculations performed by different participants. AA_s or AA_{ms} is calculated as the average of AAs of participants for the same sensitivity run (14 in our specific case). Finally, the total average amplitude (AA_t or AA_{mt}) is the average AA or average AA_m obtained from all sensitivity runs performed by all participants.

4. Application of FFT based approches to sensitivity analysis

In this section the application of the original FFTBM and FFTBM-SM is demonstrated. As was explained in Subsection 2.1, two digital signals (reference and sensitive) of the same duration are used for calculating the average amplitude, which is a measure of the influence of the sensitive parameter variation. An example of calculating AA_m is given first. Then the single value influence (based on the average amplitude) for the whole time window is given. Finally, the time dependent influence is presented for the sensitivity runs. We conclude this section with the discussion.

4.1. AA_m figure of merit example

In the example the influence of sensitive parameters on three output parameters for the interval 0-119.5 s (whole transient time) is shown for calculations of participant P-14. The participant P-14 provided for each time trend 241 samples. This means that the sampling frequency was 2 Hz. The input values for FFTBM-SM were therefore the following: $f_{fix}=2$ Hz, $f_{cut}=2$ Hz, $t_b=0$ s and $t_e=119.5$ s (time 119.5 s was selected because some users provided the last data point at a value slightly smaller than 120 s). Per sampling theorem (Eq. (1)) at least 478 data points are needed. However, the minimum number of points per FFTBM-SM is 512. Table 3 shows that requesting more samples than required has a minor influence on the results.

To get some qualitative impression on the influence of input parameters on output parameters judged by FFTBM-SM, Figure 3 shows time trends of P-14 upper plenum pressure, primary side mass and cladding temperature. Visually it may be indicated that the upper plenum pressure is the most influenced by the break flow (S-1). On the other hand, hot rod parameters (S-2, S-3, S-4 and S-5) and accumulator initial mass (S-12) do not have significant influence on upper plenum pressure (see Figures 3(a2), 3(a3), 3(a4) and 3(a5)). The primary side mass inventory is the most influenced by the gap conductivity (S-2), fuel conductivity (S-5), and the accumulator liquid mass (S-12). When comparing S-2 and S-5 calculations, one may indicate that the S-5 calculation is a bit closer to the reference calculation. On the other hand, comparing the S-2 and S-12 calculation, in the case of S-2 the difference is absolutely larger than S-12,

	Upper plenum pressure				Primary side mass				Rod surface temperature			
	$N=2^9$	$N=2^{10}$	$N=2^{11}$	$N=2^{12}$	$N=2^9$	$N=2^{10}$	$N=2^{11}$	$N=2^{12}$	$N=2^9$	$N=2^{10}$	$N=2^{11}$	$N=2^{12}$
S-1	0.085	0.088	0.089	0.089	0.102	0.103	0.103	0.104	0.330	0.335	0.338	0.339
S-2	0.032	0.033	0.034	0.034	0.124	0.125	0.125	0.125	0.799	0.806	0.805	0.807
S-3	0.015	0.016	0.016	0.016	0.056	0.056	0.057	0.057	0.488	0.491	0.494	0.495
S-4	0.015	0.016	0.016	0.016	0.056	0.057	0.057	0.057	0.295	0.301	0.301	0.303
S-5	0.026	0.027	0.027	0.027	0.108	0.108	0.109	0.109	0.937	0.944	0.948	0.948
S-6	0.019	0.019	0.020	0.020	0.046	0.046	0.046	0.046	0.256	0.258	0.261	0.263
S-7	0.023	0.024	0.024	0.024	0.051	0.051	0.052	0.052	0.258	0.269	0.270	0.272
S-8	0.016	0.016	0.017	0.017	0.041	0.041	0.042	0.042	0.143	0.145	0.147	0.148
S-9	0.018	0.019	0.019	0.019	0.039	0.040	0.040	0.040	0.205	0.208	0.209	0.210
S-10	0.015	0.016	0.016	0.017	0.043	0.043	0.044	0.044	0.449	0.460	0.460	0.461
S-11	0.021	0.022	0.022	0.022	0.090	0.090	0.091	0.091	0.239	0.240	0.243	0.245
S-12	0.029	0.030	0.030	0.031	0.117	0.118	0.119	0.119	0.480	0.487	0.488	0.490
S-13	0.045	0.046	0.046	0.047	0.065	0.066	0.066	0.066	0.302	0.308	0.310	0.311
S-14	0.015	0.016	0.016	0.016	0.052	0.053	0.053	0.053	0.255	0.256	0.259	0.261
S-15	0.016	0.016	0.017	0.017	0.059	0.059	0.060	0.060	0.147	0.150	0.151	0.152

Table 3. Average amplitude AAm for the whole time interval for participant P-14 calculated by FFFTB-SM as a function of the number of samples N

however in case of S-12 the difference is present a longer time than in the case of S-2. The calculated AA_m is comparable.

To see this in more detail, Figure 4 shows part of the magnitude difference signal spectra $|\tilde{\Delta}F(f_n)|$ for S-2 and S-12, which are used for the calculation of AA_{dif} per Eq. (5). Please note that f_{max} is 2.14 Hz. However, summing of amplitudes up to 0.2 Hz contributes more than 90% to total AA_{dif} (for S-2 the sum is 11.26 out of 12.43 and for S-12 the sum is 10.83 out of 11.79). Summing amplitudes up to 0.05 Hz (representing 13 samples out of 513) contributes more than 80% to total AA_{dif} (for S-2 the sum is 10.16 out of 12.43 and for S-12 the sum is 9.57 out of 11.79). One may see that the zero frequency component (mean value of difference signal in the time domain) is larger for S-2 than S-12 and that due to this contribution finally S-2 is judged as more influential than S-12.

4.2. Single value influence for whole time window

In the example presented in Section 4.1, AA_m for the P-14 calculation was determined. In this section all fourteen calculations are considered, and both FFTBM and FFTBM-SM are used for calculating all figures of merit presented in Section 3.3. The obtained results for the sensitive parameter influence on the three output parameters (upper plenum pressure, primary side mass inventory and rod surface temperature) are shown in Tables 4 through 6.

The qualitative comparison between FFTBM and FFTBM-SM results showed that the agreement is quite good. This is expected as at the end of the transient the influence of the sensitive parameter is generally insignificant, resulting that in the difference signal the edge is very

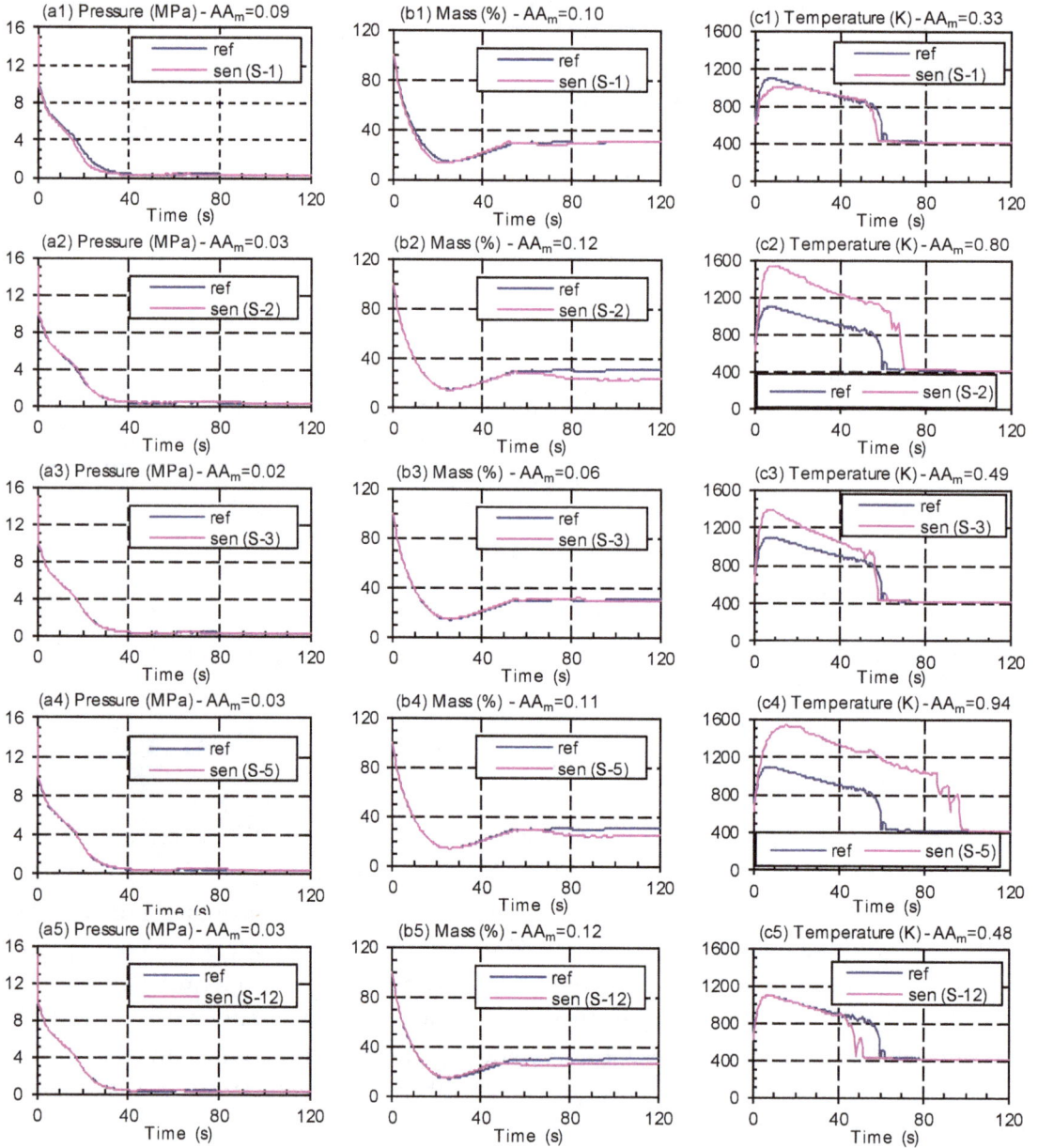

Figure 3. Participant P-14 time trends of (a) primary pressure, (b) primary side mass and (c) rod surface temperature for S-1, S-2, S-3, S-5 and S-12 with AA_m

small or not present at all. The edge is still present in the reference signal, but because it is used for the normalization it has no impact on the ranking of parameters and so the qualitative agreement between FFTBM and FFTBM-SM is good. This is not the case for the quantitative agreement as the normalization directly impacts the average amplitude. The average amplitudes obtained by both FFTBM and FFTBM-SM suggest that the most influential parameter for the upper plenum pressure is in all calculations the break flow area (S-1). To judge how influential the parameter is, the average amplitude of the sensitivity runs for the same sensitive parameter (AA_s or AA_{ms}) obtained both by FFTBM and FFTBM-SM show that the variations of the break flow area (S-1), pressurizer level (S-13), core pressure drop (S-6) and presence of crud (S-4) the most influence the output parameter upper plenum pressure. The only difference between the FFTBM and FFTBM-SM results is that ranks for S-6 and S-4 are changed.

(a) Application of FFTBM - upper plenum pressure, time interval 0 - 119.5 s

ID-S	S-1	S-2	S-3	S-4	S-5	S-6	S-7	S-8	S-9	S-10	S-11	S-12	S-13	S-14	S-15	
ID-P	AA	AA	AA	AA	AA	AA	AA	AA	AA	AA	AA	AA	AA	AA	AA	AA_p
P-1	0.051	0.011	0.010	0.006	0.009	0.011	0.008	0.013	0.011	0.019	0.019	0.015	0.036	0.011	0.010	● 0.016
P-2	0.042	0.008	0.010		0.008	0.005		0.002	0.006	0.001	0.006	0.015	0.013	0.001	0.002	● 0.009
P-3	0.070	0.025	0.035	0.030	0.034	0.108		0.045	0.028	0.042	0.043	0.045	0.060	0.010	0.008	◆ 0.042
P-4	0.110		0.014		0.013	0.015		0.011	0.011	0.012	0.030	0.033	0.027	0.010	0.012	▲ 0.025
P-5	0.089	0.014				0.016		0.023	0.018	0.019						◆ 0.030
P-6	0.029	0.010	0.012	0.013	0.010	0.077	0.015	0.009	0.011	0.010	0.011	0.010	0.017	0.008	0.005	● 0.016
P-7	0.030					0.008		0.006		0.005	0.009					● 0.012
P-8	0.068	0.013	0.009	0.010	0.012	0.015	0.018	0.014	0.015	0.015	0.016	0.018	0.027	0.012	0.010	▲ 0.018
P-9	0.074	0.007	0.009	0.008	0.008	0.010	0.009	0.009	0.010	0.008	0.011	0.015	0.027	0.000	0.008	● 0.014
P-10	0.059	0.030	0.018		0.008	0.023		0.006	0.007	0.029	0.039	0.009	0.052	0.004	0.012	▲ 0.023
P-11	0.035	0.008	0.012	0.117	0.016	0.012		0.012	0.013	0.011	0.019	0.019	0.028	0.012	0.016	▲ 0.024
P-12	0.035	0.006	0.009	0.007	0.008	0.009	0.010	0.004	0.009	0.008	0.009	0.006	0.021	0.004	0.009	● 0.010
P-13	0.041	0.014	0.015	0.011	0.016	0.132	0.016	0.014	0.013	0.015	0.017	0.011	0.024	0.006	0.005	▲ 0.023
P-14	0.044	0.017	0.008	0.008	0.013	0.010	0.012	0.008	0.009	0.008	0.011	0.015	0.023	0.008	0.008	● 0.013
AA_s	0.055	0.014	0.013	0.023	0.013	0.032	0.013	0.012	0.012	0.014	0.018	0.017	0.030	0.007	0.009	AA_t=0.019

Legend: 0.02-0.04   >0.04

(b) Application of FFTBM-SM - upper plenum pressure, time interval 0 - 119.5 s

ID-S	S-1	S-2	S-3	S-4	S-5	S-6	S-7	S-8	S-9	S-10	S-11	S-12	S-13	S-14	S-15	
ID-P	AA_m	AA_m	AA_m	AA_m	AA_m	AA_m	AA_m	AA_m	AA_m	AA_m	AA_m	AA_m	AA_m	AA_m	AA_m	AA_{mp}
P-1	0.098	0.021	0.020	0.012	0.018	0.021	0.016	0.026	0.022	0.038	0.038	0.030	0.070	0.021	0.019	● 0.031
P-2	0.080	0.016	0.018		0.014	0.009		0.003	0.011	0.002	0.012	0.029	0.026	0.002	0.004	● 0.017
P-3	0.145	0.048	0.071	0.055	0.065	0.127		0.090	0.054	0.084	0.086	0.090	0.120	0.019	0.015	◆ 0.076
P-4	0.212		0.026		0.025	0.028		0.021	0.023	0.023	0.059	0.064	0.051	0.019	0.023	▲ 0.048
P-5	0.160	0.024				0.028		0.039	0.032	0.034						◆ 0.053
P-6	0.054	0.019	0.021	0.023	0.018	0.130	0.028	0.016	0.020	0.018	0.021	0.018	0.032	0.014	0.010	● 0.029
P-7	0.056					0.016		0.010		0.008	0.016					● 0.021
P-8	0.127	0.024	0.017	0.018	0.024	0.029	0.034	0.026	0.029	0.028	0.031	0.036	0.050	0.024	0.018	▲ 0.034
P-9	0.139	0.013	0.017	0.015	0.015	0.018	0.017	0.017	0.019	0.015	0.022	0.028	0.053	0.000	0.014	● 0.027
P-10	0.118	0.060	0.035		0.015	0.043		0.012	0.013	0.056	0.073	0.017	0.098	0.009	0.024	▲ 0.044
P-11	0.068	0.017	0.024	0.234	0.030	0.023		0.024	0.025	0.023	0.036	0.036	0.052	0.024	0.033	▲ 0.046
P-12	0.065	0.011	0.017	0.014	0.016	0.017	0.019	0.008	0.016	0.014	0.018	0.010	0.039	0.007	0.017	● 0.019
P-13	0.072	0.024	0.026	0.019	0.027	0.225	0.029	0.024	0.023	0.026	0.030	0.018	0.042	0.010	0.007	▲ 0.040
P-14	0.085	0.032	0.015	0.015	0.026	0.019	0.023	0.016	0.018	0.015	0.021	0.029	0.045	0.015	0.016	● 0.026
AA_{ms}	0.106	0.026	0.026	0.045	0.024	0.052	0.024	0.024	0.023	0.027	0.036	0.034	0.056	0.014	0.017	AA_{mt}=0.036

Legend: 0.04-0.08   >0.08

Table 4. Influence of sensitive parameters on upper plenum pressure in time interval 0-119.5 s as judged by (a) original FFTBM and (b) improved FFTBM by signal mirroring

When looking calculations, the most influenced upper plenum calculation is P-3. To judge how the calculation is influenced by the sensitive parameters variation, the average amplitude of the participant sensitivity runs (AA_p or AA_{mp}) is used. Besides P-3 the P-5 calculation was also judged as the much influenced by the sensitive parameter variations. Both methods qualitatively give the same results for the average amplitude of the participant sensitivity runs. The total average amplitude (AA_t or AA_{mt}) show the overall influence of the sensitive parameters variations of all calculations on the output upper plenum pressure. The higher the value is the higher is the influence. The ratio between AA_{mt} and AA_t to be 1.88 also tells what the contribution of the edge effect is in average.

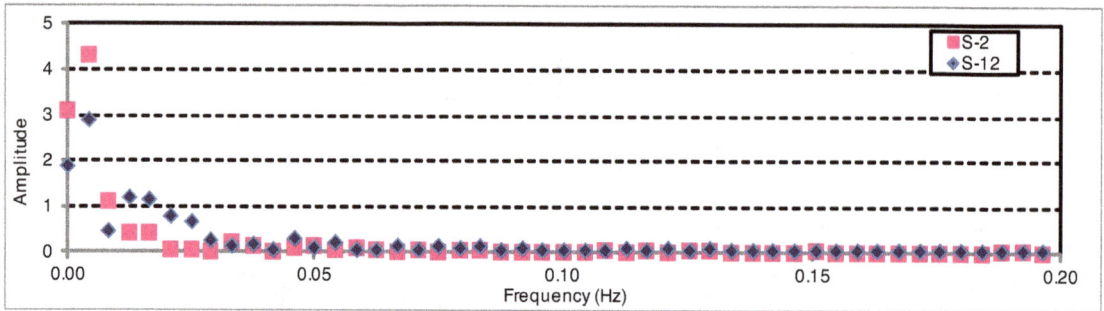

Figure 4. Magnitude difference signal spectra for S-2 and S-12 runs of P-14 participants

The average amplitudes shown in Table 5, obtained by both FFTBM and FFTBM-SM suggest that the most influential parameter for the primary side mass inventory when considering all calculations is the accumulator liquid mass (S-12). Both the AA_s and AA_{ms} indicate the accumulator liquid mass (S-12), break flow area (S-1), accumulator pressure (S-11) and pressurizer level (S-13) the most influential sensitive parameters on the primary side mass inventory.

When looking the calculations, the most influenced primary side mass inventory calculation is P-3. The AA_p and AA_{mp} indicate as the second most influenced the P-13 calculation. Again both methods qualitatively give very similar results for the average amplitude of the participant sensitivity runs. The AA_t and AA_{mt} show that the overall influence of the sensitive parameters variations of all calculations on the output primary side mass inventory is higher than on the upper plenum pressure. The ratio between AA_{mt} and AA_t is 1.55, indicating that the primary side mass inventory is less influenced by the edge effect (see also Figure3).

The average amplitudes shown in Table 6, obtained by both FFTBM and FFTBM-SM suggest that the most influential parameter for the rod surface temperature when considering all calculations is the fuel conductivity (S-5). Both AA_s and AA_{ms} indicate the fuel conductivity (S-5) and gap conductivity (S-2) as the most influential. Significant influences have also the gap thickness (S-3), maximum linear power (S-10), break flow area (S-1) and the accumulator liquid mass (S-12) and a few others. The only difference between the FFTBM and FFTBM-SM results is that the ranks for S-1 and S-12 are changed.

(a) Application of FFTBM - primary side mass inventory, time interval 0 - 119.5 s

ID-S	S-1	S-2	S-3	S-4	S-5	S-6	S-7	S-8	S-9	S-10	S-11	S-12	S-13	S-14	S-15	
ID-P	AA	AA	AA	AA	AA	AA	AA	AA	AA	AA	AA	AA	AA	AA	AA	AAp
P-1	0.052	0.026	0.019	0.018	0.015	0.026	0.018	0.017	0.015	0.017	0.048	0.052	0.032	0.014	0.02	● 0.026
P-2	0.046	0.014	0.014		0.013	0.015		0.014	0.015	0.016	0.017	0.084	0.027	0.012	0.02	● 0.023
P-3	0.102	0.083	0.071	0.060	0.092	0.180		0.118	0.064	0.086	0.124	0.137	0.124	0.013	0.01	◆ 0.091
P-4	0.125		0.022		0.021	0.042		0.018	0.037	0.023	0.054	0.091	0.036	0.025	0.03	▲ 0.043
P-5	0.071	0.027			0.022		0.051	0.038	0.033							▲ 0.040
P-6	0.066	0.036	0.056	0.020	0.052	0.045	0.061	0.035	0.035	0.036	0.034	0.158	0.044	0.040	0.07	▲ 0.052
P-7	0.035				0.035			0.017		0.007	0.026					● 0.024
P-8	0.083	0.053	0.016	0.046	0.025	0.051	0.042	0.022	0.038	0.033	0.042	0.073	0.069	0.031	0.03	▲ 0.043
P-9	0.073	0.014	0.013	0.015	0.011	0.012	0.012	0.022	0.020	0.008	0.038	0.093	0.033	0.000	0.06	● 0.028
P-10	0.089	0.010	0.003		0.059	0.019		0.039	0.034	0.019	0.153	0.095	0.041	0.038	0.04	▲ 0.049
P-11	0.041	0.019	0.024	0.145	0.041	0.020		0.048	0.020	0.033	0.040	0.053	0.032	0.039	0.07	▲ 0.045
P-12	0.039	0.010	0.033	0.018	0.022	0.022	0.024	0.007	0.037	0.025	0.024	0.203	0.039	0.029	0.09	▲ 0.042
P-13	0.072	0.047	0.035	0.045	0.066	0.077	0.000	0.050	0.077	0.044	0.092	0.179	0.075	0.049	0.07	◆ 0.065
P-14	0.049	0.136	0.035	0.034	0.104	0.032	0.026	0.022	0.023	0.023	0.049	0.090	0.060	0.041	0.05	▲ 0.052
AAs	0.067	0.040	0.028	0.044	0.043	0.043	0.026	0.034	0.035	0.029	0.057	0.109	0.051	0.028	0.05	AAt=0.046

Legend: ▓ 0.05-0.1   ▓ >0.1

(b) Application of FFTBM-SM - primary side mass inventory, time interval 0 - 119.5 s

ID-S	S-1	S-2	S-3	S-4	S-5	S-6	S-7	S-8	S-9	S-10	S-11	S-12	S-13	S-14	S-15	
ID-P	AAm	AAm	AAm	AAm	AAm	AAm	AAm	AAm	AAm	AAm	AAm	AAm	AAm	AAm	AAm	AAmp
P-1	0.096	0.043	0.034	0.030	0.028	0.034	0.030	0.033	0.028	0.032	0.071	0.084	0.062	0.025	0.037	● 0.045
P-2	0.096	0.028	0.028		0.025	0.028		0.025	0.028	0.029	0.035	0.139	0.047	0.024	0.033	● 0.044
P-3	0.171	0.146	0.142	0.107	0.158	0.215		0.212	0.124	0.168	0.202	0.226	0.229	0.027	0.027	◆ 0.154
P-4	0.227		0.041		0.040	0.076		0.036	0.063	0.045	0.107	0.160	0.071	0.043	0.048	▲ 0.080
P-5	0.141	0.049			0.041		0.086	0.072	0.063							▲ 0.076
P-6	0.096	0.058	0.076	0.035	0.070	0.074	0.081	0.050	0.055	0.055	0.060	0.159	0.085	0.055	0.069	▲ 0.072
P-7	0.060				0.051			0.020		0.014	0.048					● 0.039
P-8	0.152	0.073	0.032	0.091	0.047	0.094	0.081	0.042	0.074	0.067	0.077	0.133	0.084	0.062	0.053	▲ 0.078
P-9	0.136	0.024	0.019	0.022	0.020	0.022	0.021	0.026	0.029	0.014	0.052	0.134	0.057	0.000	0.062	● 0.043
P-10	0.111	0.017	0.006		0.073	0.025		0.048	0.043	0.034	0.161	0.101	0.065	0.062	0.062	▲ 0.062
P-11	0.070	0.038	0.044	0.290	0.064	0.038		0.077	0.035	0.058	0.068	0.091	0.060	0.057	0.094	▲ 0.077
P-12	0.057	0.011	0.035	0.019	0.025	0.025	0.030	0.007	0.037	0.027	0.035	0.135	0.050	0.019	0.066	● 0.039
P-13	0.133	0.085	0.063	0.085	0.102	0.129	0.000	0.092	0.128	0.076	0.144	0.235	0.092	0.091	0.124	◆ 0.105
P-14	0.102	0.124	0.056	0.056	0.108	0.046	0.051	0.041	0.039	0.043	0.090	0.117	0.065	0.052	0.059	▲ 0.070
AAms	0.118	0.058	0.048	0.082	0.063	0.064	0.042	0.057	0.058	0.052	0.088	0.143	0.081	0.043	0.061	AAmt=0.071

Legend: ▓ 0.07-0.14   ▓ >0.14

Table 5. Influence of sensitive parameters on primary side mass inventory in time interval 0-119.5 s as judged by (a) original FFTBM and (b) improved FFTBM by signal mirroring

When looking the calculations, the most influenced rod surface rod temperature calculation is P-10. The AA_p and AA_{mp} indicate that the next two most influenced are the P-2 and P-3 calculation. Again both methods qualitatively give pretty similar results for the average amplitude of the participant sensitivity runs. The AA_t and AA_{mt} show that the overall influence of the sensitive parameters variations of all calculations on the output rod surface temperature

(a) Application of FFTBM - rod surface temperature, time interval 0 - 119.5 s

ID-S	S-1	S-2	S-3	S-4	S-5	S-6	S-7	S-8	S-9	S-10	S-11	S-12	S-13	S-14	S-15	
ID-P	AA	AA	AA	AA	AA	AA	AA	AA	AA	AA	AA	AA	AA	AA	AA	AA_p
P-1	0.323	0.498	0.424	0.085	0.318	0.189	0.122	0.194	0.253	0.413	0.283	0.342	0.204	0.104	0.129	🟢 0.259
P-2	0.323	0.403	0.608		0.493	0.312		0.301	0.282	0.495	0.220	1.679	0.097	0.099	0.174	🔶 0.422
P-3	0.432	0.451	0.476	0.406	0.703	0.496		0.415	0.386	0.266	0.426	0.198	0.212	0.247	0.184	🔶 0.378
P-4	0.368		0.252		0.445	0.185		0.210	0.289	0.527	0.195	0.388	0.207	0.328	0.107	🔺 0.292
P-5	0.184	0.284				0.174		0.330	0.308	0.328						🟢 0.268
P-6	0.622	0.379	0.402	0.167	0.543	0.415	0.444	0.349	0.157	0.483	0.392	0.210	0.360	0.172	0.166	🔺 0.351
P-7	0.340					0.207		0.239		0.358	0.261					🔺 0.281
P-8	0.446	0.396	0.305	0.150	0.351	0.227	0.283	0.329	0.337	0.360	0.117	0.434	0.149	0.171	0.133	🔺 0.279
P-9	0.229	0.407	0.343	0.135	0.386	0.208	0.089	0.152	0.255	0.207	0.208	0.212	0.160	0.000	0.072	🟢 0.204
P-10	0.345	0.848	0.555		0.762	0.264		0.349	0.339	0.683	0.311	0.238	0.401	0.080	0.307	🔶 0.422
P-11	0.207	0.507	0.639	0.335	0.243	0.358		0.681	0.204	0.437	0.224	0.239	0.279	0.104	0.142	🔺 0.328
P-12	0.344	0.598	0.419	0.216	0.543	0.192	0.110	0.252	0.339	0.825	0.232	0.187	0.245	0.145	0.211	🔺 0.324
P-13	0.253	0.535	0.339	0.124	0.464	0.343	0.238	0.197	0.349	0.221	0.290	0.256	0.111	0.090	0.160	🟢 0.265
P-14	0.257	0.668	0.420	0.267	0.738	0.206	0.205	0.113	0.173	0.386	0.197	0.380	0.236	0.204	0.114	🔺 0.304
AA_s	0.334	0.498	0.432	0.209	0.499	0.270	0.213	0.294	0.282	0.428	0.258	0.397	0.222	0.145	0.158	AA_t=0.314

0.25-0.45　　　>0.45

(b) Application of FFTBM-SM - rod surface temperature, time interval 0 - 119.5 s

ID-S	S-1	S-2	S-3	S-4	S-5	S-6	S-7	S-8	S-9	S-10	S-11	S-12	S-13	S-14	S-15	
ID-P	AA_m	AA_m	AA_m	AA_m	AA_m	AA_m	AA_m	AA_m	AA_m	AA_m	AA_m	AA_m	AA_m	AA_m	AA_m	AA_{mp}
P-1	0.396	0.605	0.499	0.103	0.383	0.229	0.149	0.231	0.302	0.490	0.338	0.415	0.247	0.127	0.155	🟢 0.311
P-2	0.389	0.475	0.717		0.605	0.368		0.358	0.348	0.542	0.265	0.913	0.121	0.120	0.214	🔶 0.418
P-3	0.508	0.515	0.516	0.481	0.861	0.580		0.486	0.461	0.305	0.489	0.234	0.250	0.285	0.217	🔶 0.442
P-4	0.438		0.285		0.504	0.226		0.255	0.350	0.557	0.230	0.459	0.249	0.402	0.131	🔺 0.340
P-5	0.227	0.360				0.213		0.402	0.385	0.400						🔺 0.331
P-6	0.736	0.467	0.444	0.200	0.656	0.481	0.526	0.412	0.183	0.525	0.452	0.247	0.423	0.201	0.196	🔺 0.410
P-7	0.411					0.251		0.288		0.416	0.316					🔺 0.336
P-8	0.528	0.456	0.372	0.188	0.432	0.274	0.342	0.404	0.414	0.449	0.149	0.483	0.189	0.209	0.165	🔺 0.337
P-9	0.293	0.492	0.418	0.172	0.506	0.264	0.114	0.194	0.311	0.238	0.264	0.270	0.202	0.000	0.091	🟢 0.255
P-10	0.410	0.982	0.625		0.930	0.307		0.402	0.391	0.768	0.367	0.284	0.472	0.094	0.353	🔶 0.491
P-11	0.244	0.579	0.727	0.412	0.270	0.442		0.829	0.243	0.523	0.278	0.286	0.342	0.126	0.170	🔺 0.391
P-12	0.414	0.680	0.487	0.260	0.667	0.231	0.128	0.309	0.383	0.637	0.282	0.227	0.293	0.170	0.251	🔺 0.361
P-13	0.302	0.604	0.401	0.148	0.563	0.411	0.291	0.238	0.412	0.266	0.351	0.292	0.134	0.109	0.177	🟢 0.313
P-14	0.330	0.799	0.488	0.295	0.937	0.256	0.258	0.143	0.205	0.449	0.239	0.480	0.302	0.255	0.147	🔺 0.372
AA_{ms}	0.402	0.585	0.498	0.251	0.609	0.324	0.258	0.353	0.338	0.469	0.309	0.382	0.269	0.175	0.189	AA_{mt}=0.366

Legend:　　　0.3-0.5　　　>0.5

Table 6. Influence of sensitive parameters on rod surface temperature in time interval 0-119.5 s as judged by (a) original FFTBM and (b) improved FFTBM by signal mirroring

is higher than on the primary side mass inventory. The ratio between AA_{mt} and AA_t is 1.17, indicating that the rod surface temperature is the least influenced by the edge effect (see also Figure 3).

4.3. Time dependent influence

The results presented in Section 4.2 give information on the accumulated influence of sensitive parameters for the whole transient duration (single value figures of merit). Additional insight into the results is obtained from the time dependent average amplitudes for each single variation of parameters. They provide information how the influence changes during the transient progression.

Figure 5 shows the comparison between the FFTBM and FFTBM-SM results for the P-14 sensitive calculations shown in Figure 3. It is shown how the sensitive parameter influence changes during the transient progression. The judged quantitative influence in Figure 5 reflects well what is seen during the visual observation of Figure 3, in which 5 out of 15 sensitive parameter variations for the three output parameters for the P-14 calculation are shown. Please note that the FFT based approaches are especially to be used when there are several calculations (fourteen in our case) with several sensitive parameter variations (fifteen in our case) to judge the influence of the sensitive parameters in an uniform way.

When looking the output parameter upper plenum pressure, both FFTBM (excluding period when edge effect significantly contributes to average amplitude) and FFTBM-SM clearly show when during the transient the parameter was influential. For all parameters shown in Figure 5 the major influence was during the first 30 seconds when the pressure was dropping. In the case of the S-1 parameter the influence was the largest (see Figure 5(a1)), but still not extremely significant. For parameters S-2, S-3, S-5 and S-12 the total influence is small. The values of average amplitudes up to 0.03 are small. This is confirmed by Figures 3(a2), 3(a3), 3 (a4) and 3 (a5) which show that the reference and sensitive signals for the upper plenum pressure practically match each other.

When looking the output parameter primary side mass inventory, the influence of the sensitive parameters is also quite small. Parameter S-1 is the most influential in the beginning of the transient (see also Figure 3(b1)), while all other shown sensitive parameters (S-2, S-3, S-5 and S-12) become more influential later into the transient. This is in agreement with the Figures 3(b2), 3(b3), 3(b4) and 3(b5), in which the differences in the first 20 seconds are practically not visible.

Finally, when looking the output parameter rod surface temperature, the influence of the sensitive parameters is the largest among the selected output parameters as shown by the plots and the average amplitude trends. The variation of the break flow area (S-1) having the largest influence on the upper plenum pressure has a lower influence on the rod surface temperature than the sensitive parameters S-2, S-3, S-5 and S-12. This is logically as the break area size directly impacts the upper plenum pressure.

From Figure 5(c1) it can be seen that the influence of S-1 on the rod surface temperature is judged to be in the beginning of the transient and at around 60 s. When comparing the sensitive signal to the reference signal in Figure 3(c1), in the beginning for the sensitive signal a slower temperature increase with under predicted peak and earlier temperature decrease (rod quench) at around 60 s can be seen. At other times the trends are similar. In the case of S-2 the temperature is over predicted and the quench is delayed. Therefore besides the initial jump

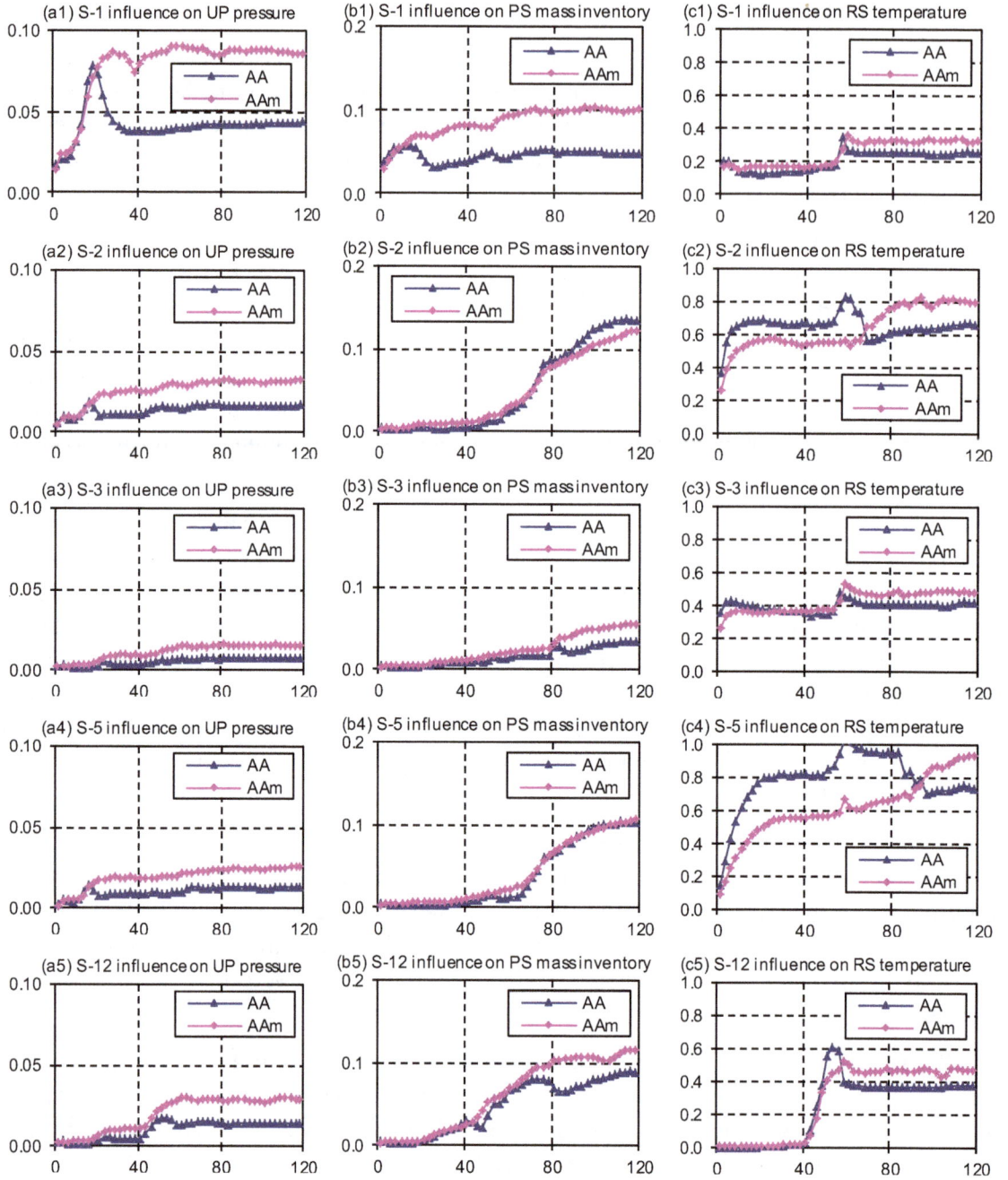

Figure 5. Participant P-14 influence of S-1, S-2, S-3, S-5, S-11 and S-12 sensitive parameter variations on (a) primary pressure and (b) primary side mass and (c) cladding temperature

the AA_m is still increasing till 20 s and for the time duration of the rod quench delay. For parameter S-3 it can be seen that its influence is between the S-1 and the S-2 influence, what can be confirmed from Figures 3(c1), 3(c2) and 3(c3). The influence of S-5 on the rod surface temperature is the largest what can be easily confirmed when comparing Figure 3(c4) and Figure 5(c4). The influence of S-11 is smaller than S-1 because S-11 influences only the time of rod quenching. Finally, S-12 shows that the larger discrepancy in the times when the rod quench starts causes also larger values of average amplitudes. Also when comparing the average amplitudes obtained by FFTBM and FFTBM-SM it can be seen that they agree pretty well for the output parameters primary side mass inventory and the rod surface temperature, because the edge present in the periodic signal is relatively smaller from the edge present in the upper plenum pressure periodic signal.

Figure 6(a) shows the comparison of the rod surface temperature reference calculations to experimental data. One may see that the calculations differ and that they do not exactly match the experimental data. Therefore the reader should always keep in mind that the direct comparison of sensitive signals (see Figure 6(b)) obtained by different participants could not answer in which calculation the sensitive parameter is the most influential.

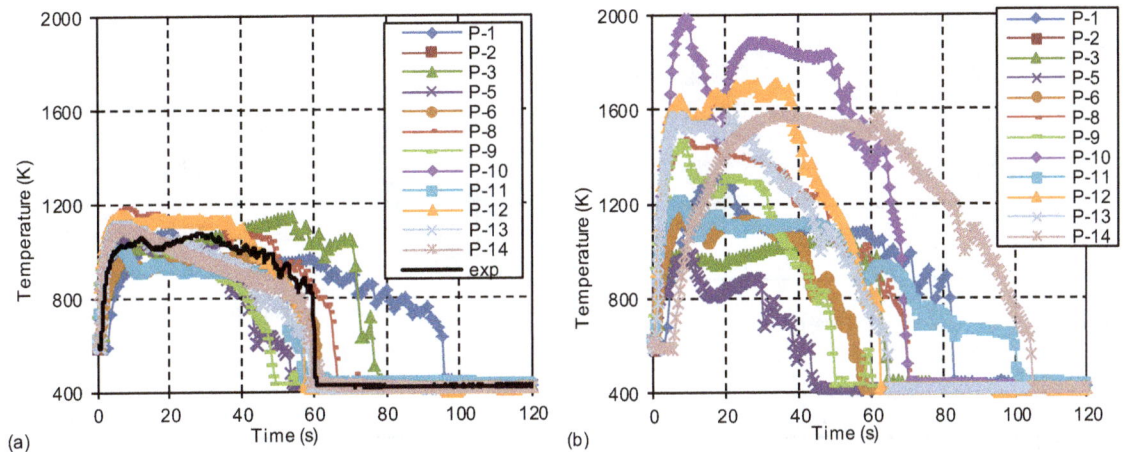

Figure 6. Comparison of (a) reference calculations with experimental value and (b) sensitive runs for S-2 sensitive parameter variation

However, by calculating the average amplitude the sensitive runs by different participants could be compared as it is shown in the Figures 7, 8 and 9 for the output parameters upper plenum pressure, primary side mass inventory and rod surface temperature, respectively. For each of the output parameters the most influential parameters are shown as identified from Tables 4, 5 and 6. Figure 7 for the upper plenum pressure shows that for the majority of calculations the influences of the same sensitive parameter variation are similar. There are only a few calculations significantly deviating, for example P-4 for S-1 variation, P-3 and P-10 for S-13 variation, P-13 for S-6 variation and P-11 for S-4 variation. Figure 8 for the primary side mass inventory shows that the P3 calculation was more sensitive to variations than other calculations. In the case of the S-4 sensitive parameter variation the P-11 calculation significantly deviates from other calculations and the reason may be that the code model is used outside its validation range [13].

Figure 9 shows that some parameters are more influential in the beginning of the transient (e.g. S-5), some in the middle of the transient (e.g. S-12) and that in the last part normally there is no significant influence (exception is the calculation P-2 for the S-12 sensitive parameter in which the rod surface temperature did not quench due to no accumulator injection when it was supposed to inject).

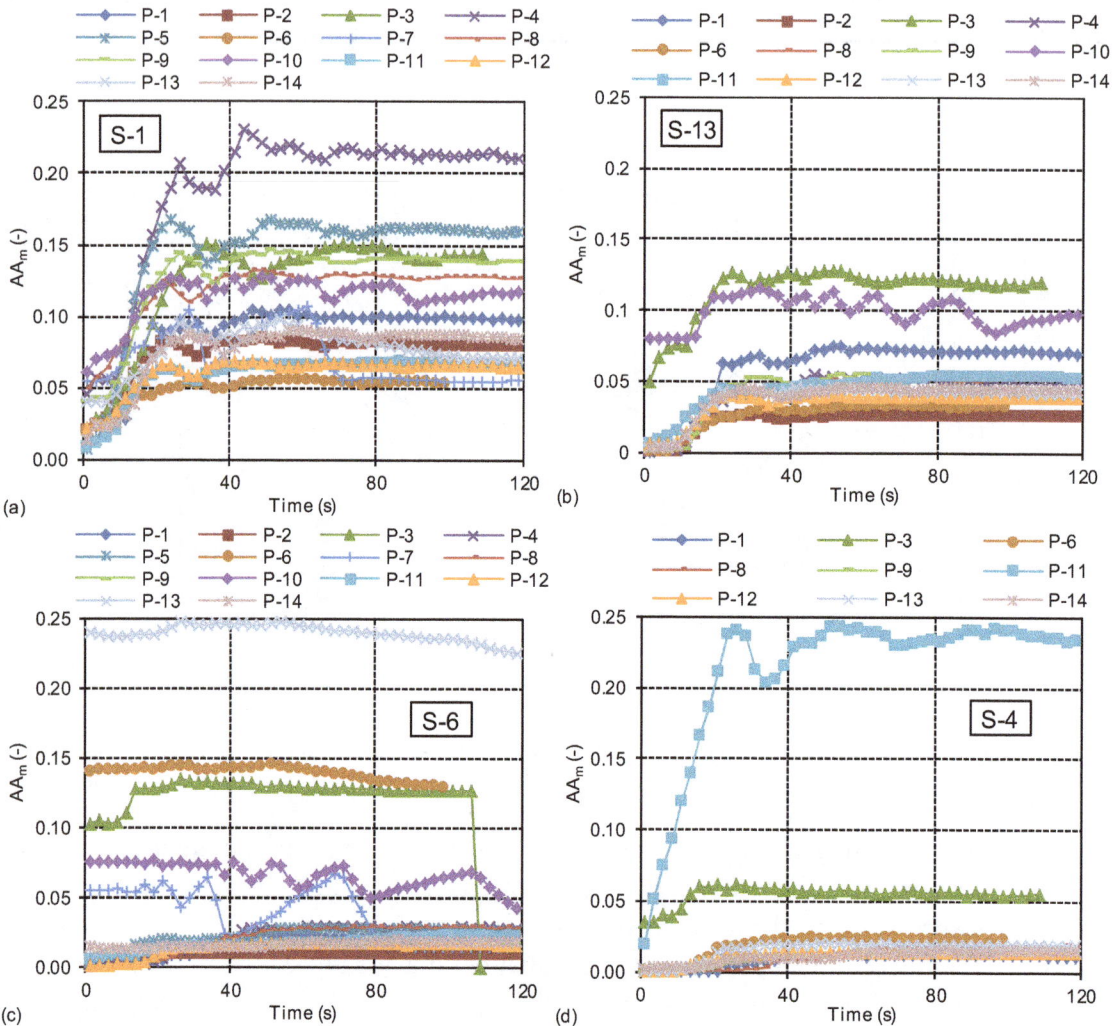

Figure 7. Upper plenum pressure time dependent AA_m of participants for (a) S-1, (b) S-13, (c) S-6 and (d) S-4 sensitive parameter variations

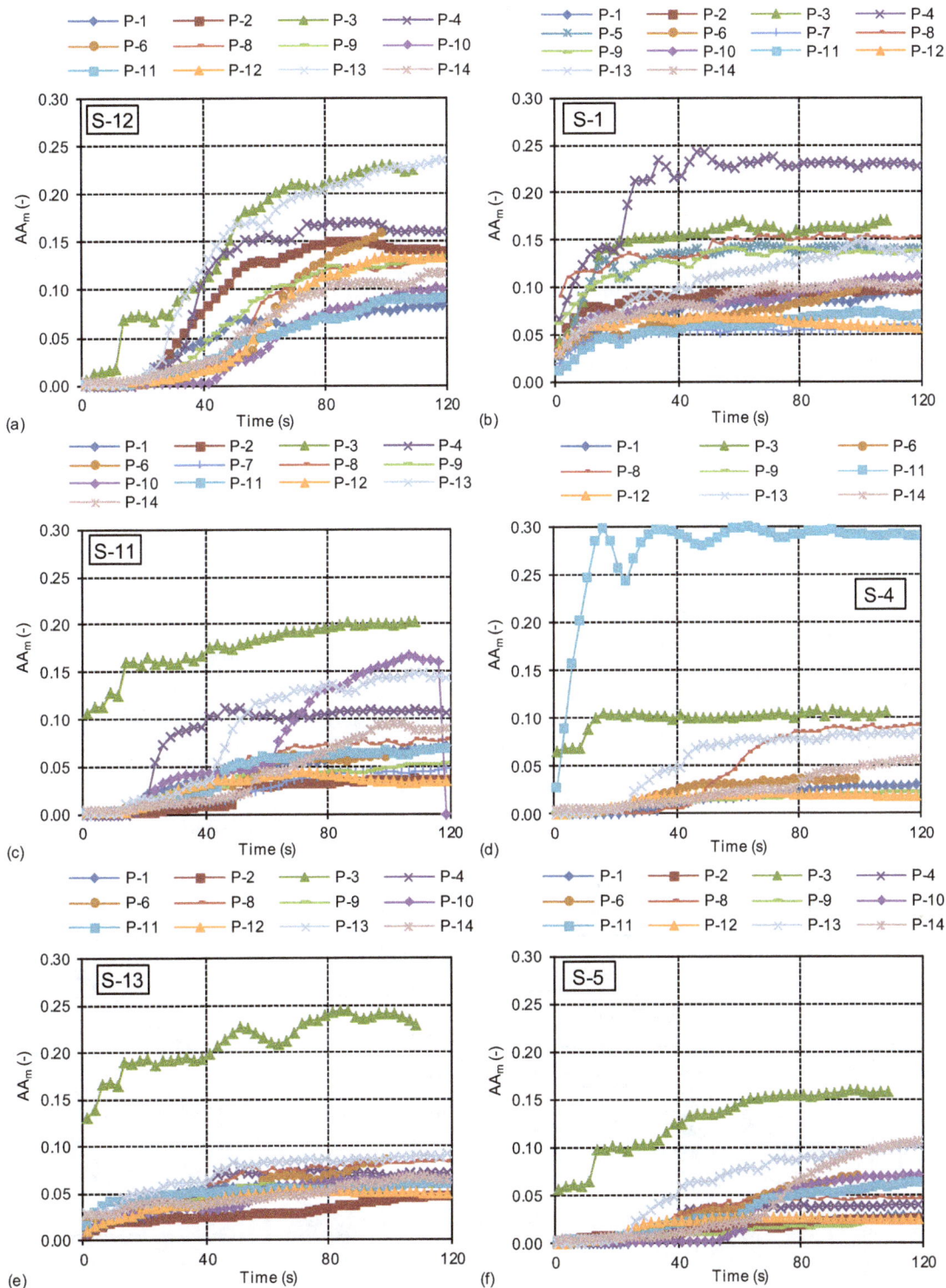

Figure 8. Primary side mass inventory time dependent AA_m of participants for (a) S-12, (b) S-1, (c) S-11, (d) S-4, (e) S-13 and (f) S-5 sensitive parameter variations

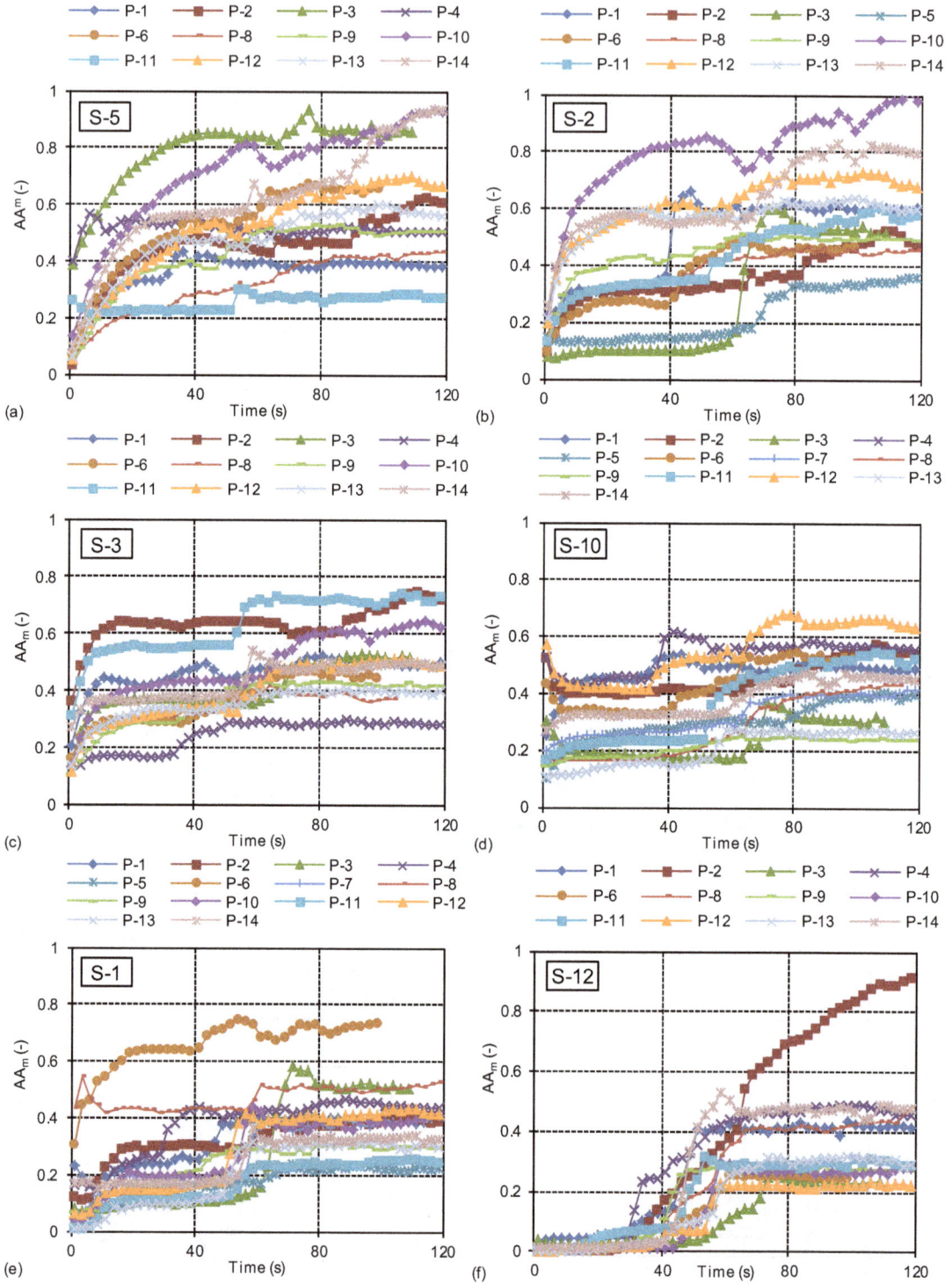

Figure 9. Rod surface temperature time dependent AA_m of participants for (a) S-5, (b) S-2, (c) S-3, (d) S-10, (e) S-1 and (f) S-12 sensitive parameter variations

4.4. Discussion

In the application for each participant each sensitivity run in the set was compared to his reference calculation and the average amplitude, which is the figure of merit for FFT based approaches, was used to judge how each output parameter is sensitive to the selected input parameter. For brevity reasons, only some examples were given in Figures 6 through 9. Nevertheless, these examples are sufficient to demonstrate that the results for the whole transient interval, shown in Tables 4, 5 and 6 may not be always sufficient for judging the sensitivity. For example, for the output parameter rod surface temperature there is higher interest in the value of the maximum rod surface temperature rather the average influence on the rod surface temperature, therefore it is important to know how the influence changes with time. Also, some parameters are more influential at the beginning and the others later in the calculation.

The results suggest that the FFT based approaches are especially appropriate for a quick sensitivity analysis in which several calculations need to be compared. It is very appropriate also due to the inherent feature, which integrates the contribution of the parameter variation with progressing transient time.

In addition, the average amplitude of participant sensitivity runs for the participants (AA_p or AA_{mp}) and the average amplitude for the same sensitive parameter (AA_s or AA_{ms}) were calculated. These measures could be used for ranking purposes. In this way information on the most influential input parameter and which participant calculation is the most sensitive to variations is obtained. Finally, different output parameters could be compared between each other regarding the influence of input parameters of all participants (AA_t or AA_{mt}). Quantitatively it is judged that the most influenced output parameter is the rod surface temperature and the least influenced the upper plenum pressure.

5. Conclusions

The study using FFTBM-SM and FFTBM was performed to show that the FFT based approaches could be used for sensitivity analyses. The LOFT L2-5 test, which simulates the large break loss of coolant accident, was used in the frame of the BEMUSE programme for sensitivity runs. In total 15 sensitivity runs were performed by 14 participants.

It can be concluded that with FFTBM-SM the analyst can get a good picture of the influence of the single parameter variation to the results throughout the transient. Some sensitive parameters are more influential at the beginning and the others later in the calculation. Due to the edge effect FFTBM-SM is advantageous for time dependent sensitivity calculations with respect to FFTBM, while for the whole transient duration (average sensitivity during whole transient) in general also FFTBM gives consistent results. FFT based approaches could be also used to quantify the influence of several parameter variations on the results. However, the influential parameters could not be identified nor the direction of the influence.

The results suggest that the FFT based approaches are especially appropriate for a quick assessment of a sensitivity analysis in which several calculations need to be compared or the influence of single sensitive parameters needs to be ranked. Such a sensitivity analysis could provide information which are the most influential parameters and how influential the input parameters are on the selected output parameters and when they influence during a transient.

Author details

Andrej Prošek and Matjaž Leskovar

Reactor Engineering Division, Jožef Stefan Institute, Ljubljana, Slovenia

References

[1] Perez M, Reventos F, Batet L, Guba A, Toth I, Mieusset T, et al. Uncertainty and sensitivity analysis of a LBLOCA in a PWR Nuclear Power Plant: Results of the Phase V of the BEMUSE programme. Nucl Eng Des. 2011;241(10):4206-4222. PubMed PMID: WOS:000295956100015. English.

[2] Prošek A, Leskovar M. Application of fast Fourier transform for accuracy evaluation of thermal-hydraulic code calculation. In: S. NG, editor. Fourier transforms - approach to scientific principles. Rijeka: In-Tech; 2011. p. 447-68.

[3] de Crecy A, Bazin P, Glaeser H, Skorek T, Joucla J, Probst P, et al. Uncertainty and sensitivity analysis of the LOFT L2-5 test: Results of the BEMUSE programme. Nucl Eng Des. 2008;238(12):3561-3578.

[4] Prošek A, Leskovar M, Mavko B. Quantitative assessment with improved fast Fourier transform based method by signal mirroring. Nucl Eng Des. 2008;238(10): 2668-2677.

[5] Prošek A, D'Auria F, Mavko B. Review of quantitative accuracy assessments with fast Fourier transform based method (FFTBM). Nucl Eng Des. 2002;217(1-2):179-206.

[6] Smith SW. The Scientist and Engineer's Guide to Digital Signal Processing. San Diego, California: California Technical Publishing, Second Edition; 1999.

[7] Lapendes DN. Dictionary of scientific and technical terms. Second Edition ed. New York, St. Louis, San Francisco, Auckland, Bogotá, Düsseldorf, Johannesburg, London, Madrid, Mexico, Montreal, New Delhi, Panama, Paris, São Paulo, Singapore, Sydney, Tokyo, Toronto, : McGraw-Hill Book Company; 1978.

[8]   Kunz RF, Kasmala GF, Mahaffy JH, Murray CJ. On the automated assessment of nuclear reactor systems code accuracy. Nucl Eng Des. 2002; 211(2-3):245-272. PubMed PMID: WOS:000174067600010. English.

[9]   Prošek A, Leskovar M. Improved FFTBM by signal mirroring as a tool for code assessment. Proceedings of the ICAPP, International congress on advances in nuclear power plants: [S. l.]: SFEN; 2007. p. 7121-1-7121-9.

[10] Prošek A, Mavko B. A tool for quantitative assessment of code calculations with an improved fast Fourier transform based method = Orodje za kvantitativno ocenitev računalniškega izračuna z metodo na podlagi hitre Fourirjeve transformacije. Electrotechnical review. 2003; 70(5):291-296.

[11] Prošek A. JSI FFTBM Add-in 2007 : users's manual. 2007. IJS work report IJS-DP-9752.

[12] Prošek A, Mavko B. Quantitative assessment of time trends : influence of time window selection. In: Pevec D, Debrecin N, editors. 5th International Conference Nuclear Option in Countries with Small and Medium Electricity Grids; 16-20 May 2004; Dubrovnik, Croatia. Zagreb: Croatian Nuclear Society; 2004.

[13] OECD/NEA. BEMUSE Phase VI Report - Status report on the area, classification of the methods, conclusions and recommendations. Paris, France: OECD/NEA, 2011 NEA/CSNI/R(2011)4.

[14] OECD/NEA. BEMUSE Phase III Report, Uncertainty and Sensitivity Analysis of the LOFT L2-5 Test. Paris, France: OECD/NEA, 2007 NEA/CSNI/R(2007)4.

[15] OECD/NEA. BEMUSE Phase II Report, Re-Analysis of the ISP-13 Exercise, Post Test Analysis of the LOFT L2-5 Test Calculation. Paris, France: OECD/NEA, 2006 NEA/CSNI/R(2006)2.

7

Jacket Matrix Based Recursive Fourier Analysis and Its Applications

Daechul Park and Moon Ho Lee

Additional information is available at the end of the chapter

1. Introduction

The last decade based on orthogonal transform has been seen a quiet revolution in digital video technology as in Moving Picture Experts Group (MPEG)-4, H.264, and high efficiency video coding (HEVC) [1–7]. The discrete cosine transform (DCT)-II is popular compression structures for MPEG-4, H.264, and HEVC, and is accepted as the best suboptimal transformation since its performance is very close to that of the statistically optimal Karhunen-Loeve transform (KLT) [1-5].

The discrete signal processing based on the discrete Fourier transform (DFT) is popular in wide range of applications depending on specific targets: orthogonal frequency division multiplexing (OFDM) wireless mobile communication systems in 3GPP-LTE [3], mobile worldwide interoperability for microwave access (WiMAX), international mobile telecommunications-advanced (IMT-Advanced), broadcasting related applications such as digital audio broadcasting (DAB), digital video broadcasting (DVB), digital multimedia broadcasting (DMB)) based on DFT. Furthermore, the Haar-based wavelet transform (HWT) is also very useful in the joint photographic experts group committee in 2000 (JPEG-2000) standard [2], [8]. Thus, different applications require different types of unitary matrices and their decompositions. From this reason, in this book chapter we will propose a unified hybrid algorithm which can be used in the mentioned several applications in different purposes.

Compared with the conventional individual matrix decompositions, our main contributions are summarized as follows:

- We propose the diagonal sparse matrix factorization for a unified hybrid algorithm based on the properties of the Jacket matrix [9], [10] and the recursive decomposition of the sparse matrix. It has been shown that this matrix decomposition is useful in developing the fast algorithms [11]. Individual DCT-II [1–3], [6], [7], [12], DST-II [4], [6], [7], [13], DFT [3], [5], [14], and HWT [8] matrices can be decomposed to one orthogonal character matrix and a corresponding special sparse matrix. The inverse of the sparse matrix can be easily obtained from the property of the block (element)-wise inverse Jacket matrix. However, there have been no previous works in the development of the common matrix decomposition supporting these transforms.

- We propose a new unified hybrid algorithm which can be used in the multimedia applications, wireless communication systems, and broadcasting systems at almost the same computational complexity as those of the conventional unitary matrix decompositions as summarized in Table 1 and 2. Compared with the existing unitary matrix decompositions, the proposed hybrid algorithm can be even used to the heterogeneous systems with hybrid multimedia terminals being serviced with different applications. The block (element)-wise diagonal decompositions of DCT-II, DST-II, DFT and DWT have a similar pattern as Cooley-Tukey's regular butterfly structures. Moreover, this unified hybrid algorithm can be also applied to the wireless communication terminals requiring a multiuser multiple input-multiple output (MIMO) SVD block diagonalization systems [15], [11,19], [22] and diagonal channels interference alignment management in macro/femto cell coexisting networks [16]. In [15-16, 19, 22- 23], a block-diagonalized matrix can be applied to wireless communications MIMO downlink channel.

In Section 2, we present recursive factorization algorithms of DCT-II, DST-II, and DFT matrix for fast computation. In Section 3, hybrid architecture is proposed for fast computations of DCT-II, DST-II, and DFT matrices. Also numerical simulations follow. The conclusion is given in Section 4.

Notation: The superscript $(\cdot)^T$ denotes transposition; I_N denotes the $N \times N$ identity matrix; 0 denotes an all-zero matrix of appropriate dimensions; $C_l^i = \cos(i\pi / l)$; $S_l^i = \sin(i\pi / l)$; $W = e^{-\frac{j2\pi}{N}}$; $\otimes$ and $\oplus$, respectively, denote the Kronecker product and the direct sum.

2. Jacket matrix based recursive decompositions of Fourier matrix

2.1. Recursive decomposition of DCT-II

Definition 1: Let $J_N = \{a_{i,j}\}$ be a matrix, then it is called the Jacket matrix when $J_N^{-1} = \frac{1}{N}\{(a_{i,j})^{-1}\}^T$.

That is, the inverse of the Jacket matrix can be determined by its element-wise inverse [9-11]. The row permutation matrix, P_N is defined by

$$P_2 = I_2 \text{ and } P_N = \begin{bmatrix} 1 & 0 & 0 & \cdots & 0 & 0 & \cdots & 0 \\ 0 & 0 & 0 & \cdots & 1 & 0 & \cdots & 0 \\ 0 & 1 & 0 & \cdots & 0 & 0 & \cdots & 0 \\ 0 & 0 & 0 & \cdots & 0 & 1 & \cdots & 0 \\ \vdots & \vdots & \vdots & \ddots & \vdots & \vdots & \ddots & \vdots \\ 0 & 0 & 0 & \cdots & 0 & 0 & \cdots & 1 \end{bmatrix}. \tag{1}$$

where P_N elements are determined by the following relation:

$$\begin{cases} p_{i,j}=1, & if \quad i=2j, \quad 0 \le j \le \dfrac{N}{2}-1, \\[2mm] p_{i,j}=1, & if \quad i=(2j+1)\bmod N, \quad \dfrac{N}{2} \le j \le N-1, \\[2mm] p_{i,j}=0, & others. \end{cases}$$

The block column permutation matrix, Q_N is defined by

$$Q_N = I_2 \text{ and } Q_N = \begin{bmatrix} I_{N/4} & 0_{N/2} \\ 0_{N/2} & \bar{I}_{N/4} \end{bmatrix}, \quad N \ge 4. \tag{2}$$

where $\bar{I}_{N/2}$ denotes reversed identity matrix. Note that $Q_N^{-1}=Q_N$ and $P_N^{-1} \ne P_N$, whereas $Q_N^{-1}=Q_N^T$ and $P_N^{-1}=P_N^T$.

Proposition 1: *With the use of the Kronecker product and Hadamard matrices, a higher order block-wise inverse Jacket matrix (BIJM) can be recursively obtained by*

$$J_{2N} = J_N \otimes H_2, N \ge 2 \tag{3}$$

then

$$J_{2N}^{-1} = \frac{1}{N} J_{2N}^T \tag{4}$$

where the lowest order Hadamard matrix is defined by $H_2 = \begin{bmatrix} 1 & 1 \\ 1 & -1 \end{bmatrix}$

Proof: *A proof of this proposition is given in Appendix 6.A.*

Note that since the BIJM requires a matrix transposition and then normalization by its size, a class of transforms can be easily inverted as follows:

$$Y_{2N} = J_{2N}X_{2N}, \text{ and } X_{2N} = J_{2N}^{-1}Y_{2N} = \frac{1}{N}J_{2N}^{T}Y_{2N}. \tag{5}$$

Due to a simple operation of the BIJM, we can reduce the complexity order as the matrix size increases. In the following, we shall use this property of the BIJM in developing a hybrid diagonal block-wise transform.

According to [1-4] and [7], the DCT-II matrix is defined as follows:

$$C_N = \sqrt{\frac{2}{N}} \begin{bmatrix} \frac{1}{\sqrt{2}} & \frac{1}{\sqrt{2}} & \cdots & \frac{1}{\sqrt{2}} \\ C_{4N}^{2k_0\Phi_0} & C_{4N}^{2k_0\Phi_1} & \cdots & C_{4N}^{2k_0\Phi_{N-1}} \\ \vdots & \vdots & \ddots & \vdots \\ C_{4N}^{2k_{N-2}\Phi_0} & C_{4N}^{2k_{N-2}\Phi_1} & \cdots & C_{4N}^{2k_{N-2}\Phi_{N-1}} \end{bmatrix} = \sqrt{\frac{2}{N}}X_N \tag{6}$$

where $\Phi_i = 2i+1$ and $k_i = i+1$. We first define a permuted DCT-II matrix $\tilde{C}_N = P_N^{-1}C_N Q_N^{-1} = \sqrt{\frac{2}{N}}P_N^{-1}X_N Q_N^{-1}$. We can readily show that the matrix X_N can be constructed recursively as follows:

$$X_N = P_N \begin{bmatrix} X_{N/2} & X_{N/2} \\ B_{N/2} & -B_{N/2} \end{bmatrix} Q_N = P_N \begin{bmatrix} X_{N/2} & 0 \\ 0 & B_{N/2} \end{bmatrix} \begin{bmatrix} I_{N/2} & I_{N/2} \\ I_{N/2} & -I_{N/2} \end{bmatrix} Q_N. \tag{7}$$

Here, the matrix B_N in (7) is given as:

$$B_N = \left\{ B_N(m,n) = C_{4N}^{f(m,n)} \right\} \tag{8}$$

where $f(m, 1) = 2m-1$ and $f(m, n+1) = f(m, n) + 2f(m, 1)$ for $m, n \in \{1, 2, \ldots, N/2\}$. For example, the matrix B_4 is given by

$$B_4 = \begin{bmatrix} C_{16}^1 & C_{16}^3 & C_{16}^5 & C_{16}^7 \\ C_{16}^3 & -C_{16}^7 & -C_{16}^1 & -C_{16}^5 \\ C_{16}^5 & -C_{16}^1 & C_{16}^7 & C_{16}^3 \\ C_{16}^7 & -C_{16}^5 & C_{16}^3 & -C_{16}^1 \end{bmatrix}. \tag{9}$$

Since $X_{N/2}^{-1} = \frac{4}{N}X_{N/2}^T$ and $B_{N/2}^{-1} = \frac{4}{N}B_{N/2}^T$, the matrix decomposition in (7) is the form of the matrix product of diagonal block-wise inverse Jacket and Hadamard matrices. The matrix $B_{N/2}$ is recursively factorized using Lemma 1.

Lemma 1:*The matrix B_N can be decomposed as:*

$$B_N = L_N X_N D_N \tag{10}$$

where a lower triangular matrix L_N is defined by $L_N = \{L_N(m, n)\}$ with elements

$$L_N(m,n) = \begin{cases} \sqrt{2}(-1)^{m-1}, \forall m \text{ and } n = 1 \\ 2(-1)^{m-1}(-1)^{n-1}, m \le n \\ 0, m > n \end{cases} \tag{11}$$

and a diagonal matrix D_N is defined by $D_N = diag\{C_{4N}^{\Phi_0}, C_{4N}^{\Phi_1}, \ldots, C_{4N}^{\Phi_{N-1}}\}$.

Proof:*A proof of this Lemma is provided in Appendix 6.B.*

Using (10), we first rewrite (7) as

$$X_N = P_N \begin{bmatrix} X_{N/2} & 0 \\ 0 & L_{N/2}X_{N/2}D_{N/2} \end{bmatrix} \begin{bmatrix} I_{N/2} & I_{N/2} \\ I_{N/2} & -I_{N/2} \end{bmatrix} Q_N$$
$$= P_N \begin{bmatrix} I_{N/2} & 0 \\ 0 & L_{N/2} \end{bmatrix} [I_2 \otimes X_{N/2}] \begin{bmatrix} I_{N/2} & 0 \\ 0 & D_{N/2} \end{bmatrix} \begin{bmatrix} I_{N/2} & I_{N/2} \\ I_{N/2} & -I_{N/2} \end{bmatrix} Q_N \tag{12}$$

which can be evaluated recursively as follows:

$$X_N = P_N \begin{bmatrix} I_{N/2} & 0 \\ 0 & L_{N/2} \end{bmatrix} \times \left[I_2 \otimes \left[\cdots \left[I_2 \otimes \underbrace{\left[P_4 \begin{bmatrix} I_2 & 0 \\ 0 & L_2 \end{bmatrix} [I_2 \otimes X_2] \begin{bmatrix} I_2 & 0 \\ 0 & D_2 \end{bmatrix} \begin{bmatrix} I_2 & I_2 \\ I_2 & -I_2 \end{bmatrix} Q_4}_{X_4} \right] \cdots \right]}_{X_{N/2}} \right]$$
$$\times \begin{bmatrix} I_{N/2} & 0 \\ 0 & D_{N/2} \end{bmatrix} \begin{bmatrix} I_{N/2} & I_{N/2} \\ I_{N/2} & -I_{N/2} \end{bmatrix} Q_N. \tag{13}$$

Note that in (13) a 2×2 Hadamard matrix is defined by $X_2 = \begin{bmatrix} 1 & 1 \\ 1 & -1 \end{bmatrix}$. Also, applying the Kronecker product of I_2 and X_4, X_8 can be obtained. Keep applying the Kronecker product of I_2 and $X_{N/2}$, the final equivalent form of X_N is obtained. Thus, the proposed systematic decomposition is based on the Jacket and Hadamard matrices.

In [17], the author proposed a recursive decimation-in-frequency algorithm, where the same decomposition specified in (10) was used. However, due to using a different permutation matrix, a different recursive form was obtained. Different recursive decomposition was proposed in [18]. Four different matrices, such as the first matrix, the last matrix, the odd numbered matrix, and the even number matrix, were proposed. Compared to the decomposition in [18], the proposed decomposition is seen to be more systematic and requires less numbers of additions and multiplications. We show a complexity comparison among the proposed decomposition and other methods in Table 1-2.

Reference number	Conventional methods		Proposed	
	Addition	Multiplication	Addition	Multiplication
W. H. Chen at el [18] DCT-II	$3N/2(\log_2 N-1)+2$	$N\log_2 N-(3N/2)+4$	$N\log_2 N$	$N/2(\log_2 N+1)$
Z. Wang[13] DST-II	$N\left(\frac{7}{4}\log_2(N)-2\right)+3$	$N\left(\frac{3}{4}\log_2(N)-1\right)+3$	$N\log_2 N$	$N/2(\log_2 N+1)$
Cooley and Tukey [21] DFT	$N\log_2 N$	$(N/2)\log_2 N$	$N\log_2 N$	$(N/2)\log_2 N$

Table 1. The comparison of computation complexity of conventional independent the DCT-II, DST-II, DFT, and hybrid DCT-II/DST-II/DFT

	Matrix Size, N	Conventional		Proposed	
		Addition	Multiplication	Addition	Multiplication
DCT-II	4	8	6	8	6
	8	26	16	24	16
	16	74	44	64	40
	32	194	116	160	96
	64	482	292	384	224
	128	1154	708	896	512
	256	2690	1668	2048	1152
DST-II	4	9	5	8	6
	8	29	13	24	16
	16	83	35	64	40
	32	219	91	160	96
	64	547	227	384	224
	128	1315	547	896	512
	256	3075	1283	2048	1152
DFT	4	8	4	8	4
	8	24	12	24	12
	16	64	32	64	32

Matrix	Conventional		Proposed	
Size, N	Addition	Multiplication	Addition	Multiplication
32	160	80	160	80
64	384	192	384	192
128	896	448	896	448
256	2048	1024	2048	1024

Table 2. Computational Complexity: DCT-II/DST-II/DFT

Applying (13), we can readily compute $C_N = \sqrt{\frac{2}{N}} X_N$. The inverse of C_N can be obtained from the properties of the sparse Jacket matrix inverse:

$$(C_N)^{-1} = \sqrt{\frac{N}{2}} (Q_N)^{-1} \begin{bmatrix} I_{N/2} & I_{N/2} \\ I_{N/2} & -I_{N/2} \end{bmatrix}^{-1} \begin{bmatrix} X_{N/2}^{-1} & 0 \\ 0 & B_{N/2}^{-1} \end{bmatrix} P_N^{-1}$$

$$= \sqrt{\frac{N}{2}} Q_N \begin{bmatrix} I_{N/2} & I_{N/2} \\ I_{N/2} & -I_{N/2} \end{bmatrix} \begin{bmatrix} X_{N/2}^T & 0 \\ 0 & B_{N/2}^T \end{bmatrix} P_N^T. \tag{14}$$

The corresponding butterfly data flow diagram of C_N is given in Fig. 1.

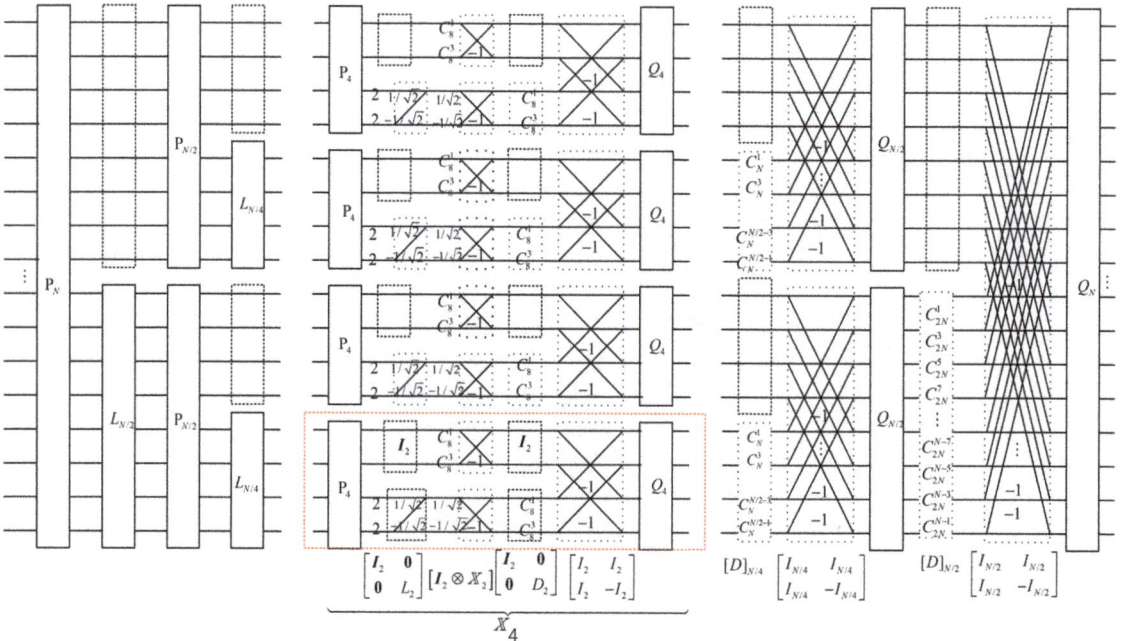

Figure 1. Regular systematic butterfly data flow of DCT-II.

2.2. Recursive decomposition of the DST-II

The DST-II matrix [1-4] and [7] can be expressed as follows:

$$
S_N = \sqrt{\frac{2}{N}}
\begin{bmatrix}
S_{4N}^{2k_0\Phi_0} & S_{4N}^{2k_0\Phi_1} & \cdots & S_{4N}^{2k_0\Phi_{N-1}} \\
S_{4N}^{2k_1\Phi_0} & S_{4N}^{2k_1\Phi_1} & \cdots & S_{4N}^{2k_1\Phi_{N-1}} \\
\vdots & \vdots & \ddots & \vdots \\
S_{4N}^{2k_{N-2}\Phi_0} & S_{4N}^{2k_{N-2}\Phi_1} & \cdots & S_{4N}^{2k_{N-2}\Phi_{N-1}} \\
\dfrac{1}{\sqrt{2}} & -\dfrac{1}{\sqrt{2}} & \cdots & -\dfrac{1}{\sqrt{2}}
\end{bmatrix}
= \sqrt{\frac{2}{N}} Y_N.
\tag{15}
$$

Similar to the procedure we have used in the DCT-II matrix, we first define the permuted DST-II matrix, $\tilde{S}_N$ as follows:

$$
\tilde{S}_N = P_N^{-1} S_N Q_N^{-1} = \sqrt{\frac{2}{N}} P_N^{-1} Y_N Q_N^{-1}.
\tag{16}
$$

From (16), we can have a recursive form for Y_N as

$$
Y_N = P_N
\begin{bmatrix}
A_{N/2} & 0 \\
0 & Y_{N/2}
\end{bmatrix}
\begin{bmatrix}
I_{N/2} & I_{N/2} \\
I_{N/2} & -I_{N/2}
\end{bmatrix}
Q_N
\tag{17}
$$

where the submatrix A_N can be calculated by

$$
A_N = U_N Y_N D_N
\tag{18}
$$

where U_N and D_N are, respectively, upper triangular and diagonal matrices. The upper triangular matrix $U_N = \{U_N(m, n)\}$ is defined as follows:

$$
U_N(m,n) =
\begin{cases}
\sqrt{2}(-1)^{m-1}, \forall m \text{ and } n = N \\
2(-1)^{m-1}(-1)^{n-1}, m \geq n \\
0, m < n
\end{cases}
\tag{19}
$$

whereas the matrix D_N is defined as before in (10). The derivation of (18) is given in Appendix C. Recursively applying (18) in (17), Recursively applying (18) in (17), we can find that

$$
\begin{aligned}
Y_N &= P_N \begin{bmatrix} A_{N/2} & 0 \\ 0 & Y_{N/2} \end{bmatrix} \begin{bmatrix} I_{N/2} & I_{N/2} \\ I_{N/2} & -I_{N/2} \end{bmatrix} Q_N = P_N \begin{bmatrix} U_{N/2}Y_{N/2}D_{N/2} & 0 \\ 0 & Y_{N/2} \end{bmatrix} \begin{bmatrix} I_{N/2} & I_{N/2} \\ I_{N/2} & -I_{N/2} \end{bmatrix} Q_N \\
&= P_N \begin{bmatrix} U_{N/2} & 0 \\ 0 & I_{N/2} \end{bmatrix} [I_2 \otimes Y_{N/2}] \begin{bmatrix} D_{N/2} & 0 \\ 0 & I_{N/2} \end{bmatrix} \begin{bmatrix} I_{N/2} & I_{N/2} \\ I_{N/2} & -I_{N/2} \end{bmatrix} Q_N.
\end{aligned}
\tag{20}
$$

Further applying (17) to the Kronecker product $[I_2 \otimes Y_{N/2}]$, the following general recursive form for DST-II matrix can be obtained as:

$$
\begin{aligned}
Y_N &= \sqrt{\frac{2}{N}} P_N \begin{bmatrix} U_{N/2} & 0 \\ 0 & I_{N/2} \end{bmatrix} \times \left[I_2 \otimes \left[\cdots \underbrace{\left[I_2 \otimes \underbrace{\left[P_4 \begin{bmatrix} U_2 & 0 \\ 0 & I_2 \end{bmatrix} [I_2 \otimes Y_2] \begin{bmatrix} D_2 & 0 \\ 0 & I_2 \end{bmatrix} \begin{bmatrix} I_2 & I_2 \\ I_2 & -I_2 \end{bmatrix} Q_4 \right]}_{Y_4} \right]}_{Y_{N/2}} \cdots \right] \right] \\
&\times \begin{bmatrix} D_{N/2} & 0 \\ 0 & I_{N/2} \end{bmatrix} \begin{bmatrix} I_{N/2} & I_{N/2} \\ I_{N/2} & -I_{N/2} \end{bmatrix} Q_N.
\end{aligned}
\tag{21}
$$

Note that if we compare (21) and (13), a similarity can be found in the proposed matrix decompositions. That is, starting from the common lowest order $Y_2 = \begin{bmatrix} 1 & 1 \\ 1 & -1 \end{bmatrix}$, the discrete sine kernel matrix is recursively constructed. Especially, applying the relationship of $U_N = \tilde{I}_N \overline{L}_N \tilde{I}_N$, where $\tilde{I}_N = \begin{bmatrix} 0 & \cdots & 0 & 1 \\ 0 & \cdots & 1 & 0 \\ \vdots & \reflectbox{$\ddots$} & \vdots & \vdots \\ 1 & \cdots & 0 & 0 \end{bmatrix}$ denotes the opposite diagonal identity matrix, the butterfly data flow of the DST-II matrix can be obtained from the corresponding that of the proposed DCT-II decomposition. The butterfly data flow graph of the DST-II matrix is shown in Fig. 2.

Now utilizing the properties of the BIJM, we can first obtain

$$
\begin{bmatrix} A_{N/2} & 0 \\ 0 & Y_{N/2} \end{bmatrix}^{-1} = \frac{2}{N} \begin{bmatrix} A_{N/2}^T & 0 \\ 0 & Y_{N/2}^T \end{bmatrix}
\tag{22}
$$

such that the inverse of the matrix S_N is given by

$$
S_N^{-1} = \sqrt{\frac{N}{2}} Q_N \begin{bmatrix} I_{N/2} & I_{N/2} \\ I_{N/2} & -I_{N/2} \end{bmatrix} \begin{bmatrix} A_{N/2}^T & 0 \\ 0 & Y_{N/2}^T \end{bmatrix} P_N^T.
\tag{23}
$$

Note that applying again the properties of the BIJM and (18), a recursive form of the inverse DST-II can be easily obtained.

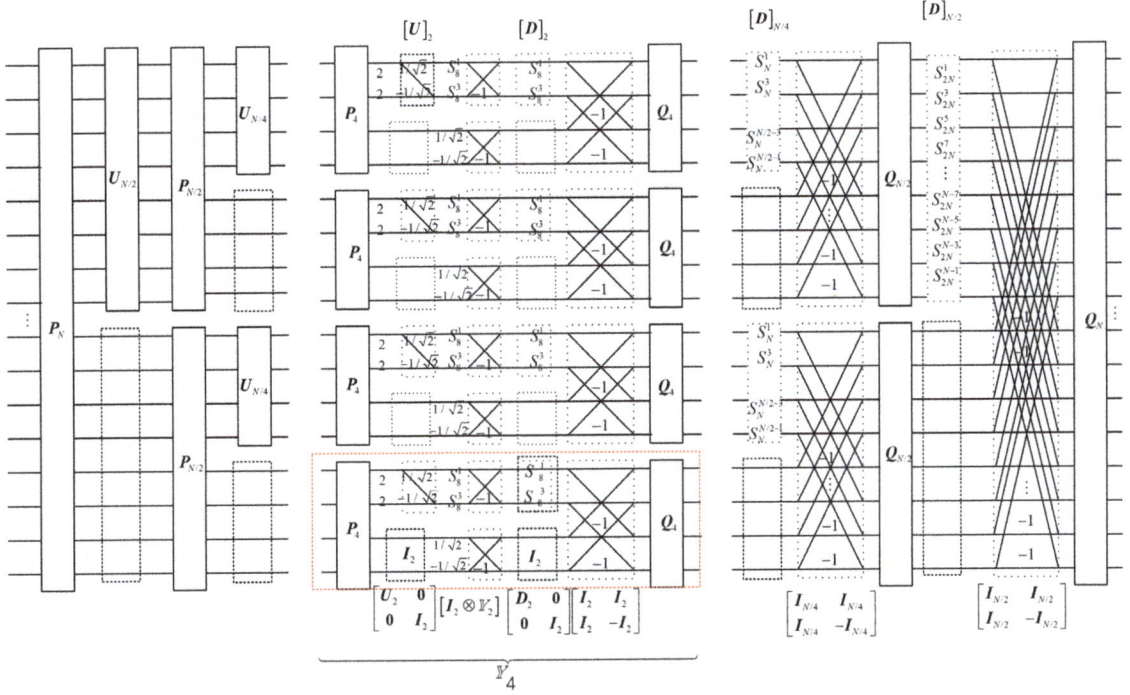

Figure 2. Regular systematic butterfly data flow of DST-II.

2.3. Recursive decomposition of DFT

The DFT is a Fourier representation of a given sequence $\{x(n)\}$,

$$X(n) = \sum_{m=0}^{N-1} x(m)W^{nm}, \quad 0 \le n \le N-1. \tag{24}$$

where $W = e^{-j2\pi/N}$. The N-point DFT matrix can be denoted by $F_N = \{W^{nm}\}$. The $N \times N$ Sylvester Hadamard matrix is denoted by H_N. The Sylvester Hadamard matrix is generated by the successive Kronecker products:

$$H_N = H_2 \otimes H_{N/2} \tag{25}$$

for $N = 4, 8, \ldots$ In (25), we define $H_2 = \begin{bmatrix} 1 & 1 \\ 1 & -1 \end{bmatrix}$. We decompose a sparse matrix $E_N = P_N \tilde{F}_N W_N$ in the following way:

$$F_N = \left[P_N\right]^T \tilde{F}_N$$

$$\tilde{F}_N = \begin{bmatrix} \tilde{F}_{N/2} & \tilde{F}_{N/2} \\ E_{N/2} & -E_{N/2} \end{bmatrix} = \begin{bmatrix} \tilde{F}_{N/2} & 0 \\ 0 & E_{N/2} \end{bmatrix} \begin{bmatrix} I_{N/2} & I_{N/2} \\ I_{N/2} & -I_{N/2} \end{bmatrix} \qquad (26)$$

where $E_{N/2}$ is further decomposed by Lemma 1

$$E_{N/2} = P_{N/2} \tilde{F}_{N/2} W_{N/2} \qquad (27)$$

where W_N is the diagonal complex unit for the N-point DFT matrix. That is, we have $W_N = diag\{W^0, \dots, W^{N-1}\}$.

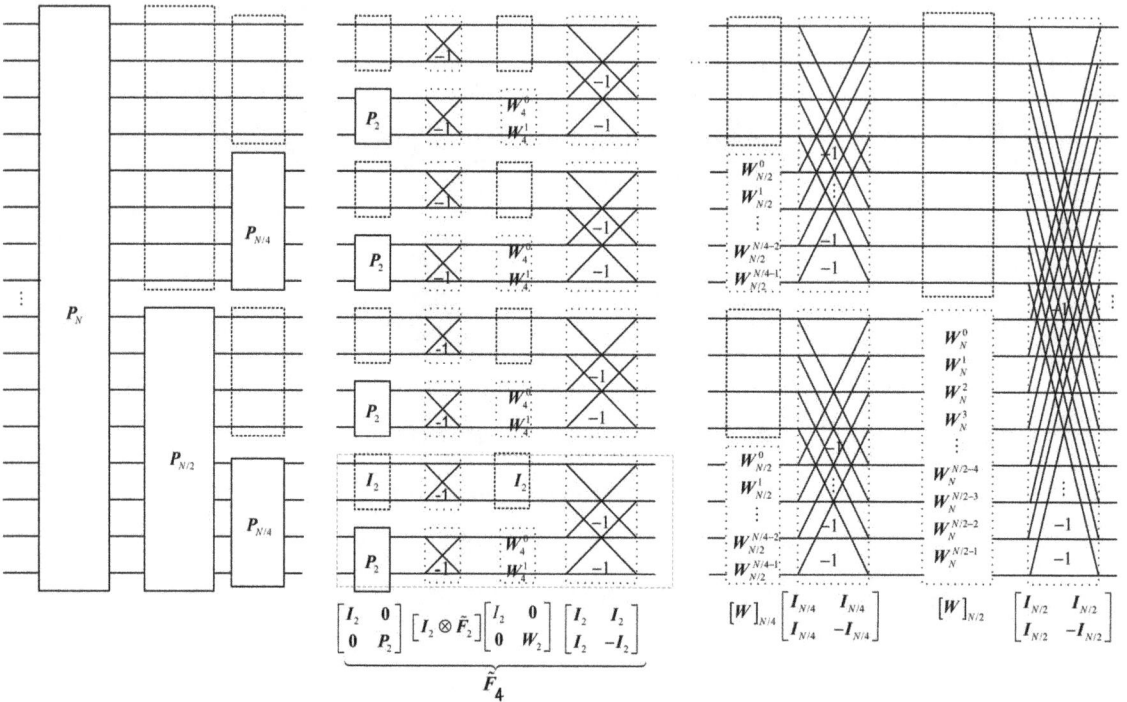

Figure 3. Butterfly data flow of DFT.

Similar to the development for DCT-II and DST-II, we first rewrite (26) using (27) as

$$\tilde{F}_N = \begin{bmatrix} \tilde{F}_{N/2} & 0 \\ 0 & P_{N/2}\tilde{F}_{N/2}W_{N/2} \end{bmatrix} \begin{bmatrix} I_{N/2} & I_{N/2} \\ I_{N/2} & -I_{N/2} \end{bmatrix}$$

$$= \begin{bmatrix} I_{N/2} & 0 \\ 0 & P_{N/2} \end{bmatrix} \begin{bmatrix} I_2 \otimes \tilde{F}_{N/2} \end{bmatrix} \begin{bmatrix} I_{N/2} & 0 \\ 0 & W_{N/2} \end{bmatrix} \begin{bmatrix} I_{N/2} & I_{N/2} \\ I_{N/2} & -I_{N/2} \end{bmatrix}. \qquad (28)$$

$\left[I_2 \otimes \tilde{F}_{N/2} \right]$ in (28) can be recursively decomposed in the following way:

$$\tilde{F}_N = \begin{bmatrix} I_{N/2} & 0 \\ 0 & P_{N/2} \end{bmatrix} \times \left[I_2 \otimes \underbrace{\left[\cdots P_4 \underbrace{\begin{bmatrix} I_2 & 0 \\ 0 & P_2 \end{bmatrix} \left[I_2 \otimes \tilde{F}_2 \right] \begin{bmatrix} I_2 & 0 \\ 0 & W_2 \end{bmatrix} \begin{bmatrix} I_2 & I_2 \\ I_2 & -I_2 \end{bmatrix}}_{\tilde{F}_4} \cdots \right]}_{\tilde{F}_{N/2}} \right] \times \begin{bmatrix} I_{N/2} & 0 \\ 0 & W_{N/2} \end{bmatrix} \begin{bmatrix} I_{N/2} & I_{N/2} \\ I_{N/2} & -I_{N/2} \end{bmatrix}. \qquad (29)$$

It is clear that the form of (29) is the same as that of (13), where we only need to change L_l to P_l and D_l to W_l for $l \in \{2, 4, 8, \ldots, N/2\}$ to convert the DCT-II matrix into DFT matrix. Consequently, the butterfly data flow of the DFT matrix can be drawn in Fig. 3 using the baseline architecture of DCT-II.

3. Proposed hybrid architecture for fast computations of DCT-II, DST-II, and DFT matrices

We have derived recursive formulas for DCT-II, DST-II, and DFT. The derived results show that DCT-II, DST-II, and DFT matrices can be unified by using a similar sparse matrix decomposition algorithm, which is based on the block-wise Jacket matrix and diagonal recursive architecture with different characters. The conventional method is only converted from DFT to DCT-II, DST-II. But our proposed method can be universally switching from DCT-II to DST-II, and DFT vice versa. Figs. 1-3 exhibit the similar recursive flow diagrams and let us motivate to develop universal hybrid architecture via switching mode selection. Moreover, the butterfly data flow graphs have $\log_2 N$ stages. From Fig.1, we can generate Figs. 2-3 according to the following proposed ways:

3.1. From DCT-II to DST-II

The N-point DCT-II of x is given by

$$X_N^{DCT}(m) = c_m \sqrt{\frac{2}{N}} \sum_{n=0}^{N-1} x(n) \cos \frac{m(2n+1)\pi}{2N} = c_m \sqrt{\frac{2}{N}} C_N \mathbf{x}$$

$$\text{where} \quad m, n = 0, 1, \ldots, N-1, \quad c_m = \begin{cases} 1 & , m \neq 0 \\ 1/\sqrt{2} & , m = 0 \end{cases} \qquad (30)$$

The N-point DST-II of **x** is given by

$$X_N^{DST}(m) = s_m \sqrt{\frac{2}{N}} \sum_{n=0}^{N-1} x(n) \sin \frac{(m+1)(2n+1)\pi}{2N} = s_m \sqrt{\frac{2}{N}} S_N \mathbf{x}$$

$$where \ m,n = 0,1,...,N-1, \ s_m = \begin{cases} 1 & ,m \neq N-1 \\ 1/\sqrt{2} & ,m = N-1 \end{cases} \tag{31}$$

Let C_N and S_N be orthogonal $N \times N$ DCT-II and DST-II matrices, respectively. Also, $\mathbf{x} = [x(0) \ x(1) \ ... \ x(N-1)]^T$ denotes the column vector for the data sequence x(n). Substituting $m = N-k-1$, $k = 1, 2, ..., N$ into (30), we have

$$C_N(N-k-1) = c_{N-k} \sqrt{\frac{2}{N}} \sum_{n=0}^{N-1} x(n) \cos \frac{(2n+1)(N-k-1)\pi}{2N}, \ k = 0,1,2,...,N-1 \tag{32}$$

Using the following trigonometric identity

$$\cos \left(\frac{(2n+1)\pi}{2} - \frac{(2n+1)(k+1)\pi}{2N} \right)$$
$$= \cos \left(\frac{(2n+1)\pi}{2} \right) \cos \left(\frac{(2n+1)(k+1)\pi}{2N} \right) + \sin \left(\frac{(2n+1)\pi}{2} \right) \sin \left(\frac{(2n+1)(k+1)\pi}{2N} \right) \tag{33}$$
$$= (-1)^n \sin \left(\frac{(2n+1)(k+1)\pi}{2N} \right)$$

(32) becomes

$$C_N(N-k-1) = c_{N-k} \sqrt{\frac{2}{N}} \sum_{n=0}^{N-1} (-1)^n x(n) \sin \frac{(2n+1)(k+1)\pi}{2N} \tag{34}$$

where $C_N = (N-k-1)$ represents the reflected version of $C_N(k)$ and this can be achieved by multiplying the reversed identity matrix $\bar{I}_N$ to C_N. (34) can be represented in a more compact matrix multiplication form [13]:

$$S_N = \bar{I}_N C_N M_N \Leftrightarrow C_N = \bar{I}_N S_N M_N \tag{35}$$

where, $M_N = [M_1 \otimes I_{N/2}]$, $M_1 = \begin{bmatrix} 1 & 0 \\ 0 & -1 \end{bmatrix}$

Then, the DST-II matrix is resulted from the DCT-II matrix. Note that compatibility property exists in the DCT-II and DST-II.

3.2. From DFT to DCT-II

The (m,n) elements of the DCT-II kernel matrix is expressed by

$$[C_N]_{m,n} = c_m \sqrt{\frac{2}{N}} \cos \frac{m(2n+1)\pi}{2N} \tag{36}$$

A new sequence $x^{(1)}(n)$ is defined by

$$\begin{cases} x^{(1)}(n) = x(2n) & for\ 0 \le n \le N/2 - 1 \\ x^{(1)}(N-n-1) = x(2n+1) & for\ 0 \le n \le N/2 - 1 \end{cases} \tag{37}$$

For the sequence $x^{(1)}(n)$, we see that we can write

$$\begin{aligned} X_N^{DCT}(m) &= c_m \sqrt{\frac{2}{N}} \sum_{n=0}^{N-1} x^{(1)}(n) \cos \frac{m(4n+1)\pi}{2N} = c_m \sqrt{\frac{2}{N}} \sum_{n=0}^{N-1} x^{(1)}(n) \cos 2\pi \frac{m}{2N} \left(2n + \frac{1}{2}\right) \\ &= c_m \sqrt{\frac{2}{N}} \mathsf{R} \left(\sum_{n=0}^{N-1} x^{(1)}(n) e^{-j2\pi m(2n+1/2)/2N} \right) = c_m \sqrt{\frac{2}{N}} \mathsf{R} \left(e^{-j\pi m/2N} \sum_{n=0}^{N-1} x^{(1)}(n) e^{-j2\pi mn/N} \right) \\ &= c_m \sqrt{\frac{2}{N}} \mathsf{R} \left(e^{-j\pi m/2N} F_N \mathbf{x}^{(1)} \right) \end{aligned} \tag{38}$$

where **R** indicates a real part.

With the result above we have avoided computing a DFT of double size. We have

$$\mathbf{W}_{4N} = diag\left\{W_{4N}{}^0, \dots, W^{N-1}\right\} = diag\left\{1, e^{-j\pi/2N}, e^{-j\pi 2/2N}, \cdots, e^{-j\pi(N-1)/2N}\right\} \tag{39}$$

Now, the result can be put in the more compact matrix-vector form

$$C_N = c_m \sqrt{\frac{2}{N}} \mathbf{R} \left(\mathbf{W}_{4N} F_N \right) \tag{40}$$

Then, the DCT-II matrix is resulted from the DFT matrix.

3.3. From DCT-II and DST-II to DFT

We develop a relation between the circular convolution operation in the discrete cosine and sine transform domains. We need to measure half of the total coefficients. The main advantage of a proposed new relation is that the input sequences to be convolved need not be symmetrical or asymmetrical. Thus, the transform coefficients can be either symmetric or asymmetric [21].

From (30) and (31), it changes to coefficient for circular convolution (C) format. Thus, we have the following equations:

$$
\mathbf{X}_N^{DCT-IIC}(m) = 2\sum_{n=0}^{N-1} x(n)\cos\left(\frac{m(2n+1)}{2N}\right), \quad m = 0,1,\cdots,N-1
$$

$$
\mathbf{X}_N^{DST-IIC}(m) = 2\sum_{n=0}^{N-1} x(n)\sin\left(\frac{m(2n+1)}{2N}\right), \quad m = 1,\cdots,N
$$

(41)

We can rewrite the DFT (24)

$$
X(m) = \sum_{n=0}^{N-1} x(n)e^{-j2\pi mn/N}, \quad m = 0,1,\ldots\ldots,N-1.
$$

(42)

Multiplying (42) by $2e^{-j\pi m/N}$, we can get

$$
2e^{-j\pi m/N}X(m) = 2e^{-j\pi m/N}\sum_{n=0}^{N-1} x(n)e^{-j2\pi mn/N} = 2\sum_{n=0}^{N-1} x(n)e^{-j\pi m/N}e^{-j2\pi mn/N}
$$

$$
= 2\sum_{n=0}^{N-1} x(n)e^{-j\left[\frac{m(2n+1)\pi}{N}\right]} = 2\sum_{n=0}^{N-1} x(n)\left(\cos\left[\frac{m(2n+1)\pi}{N}\right] - j\sin\left[\frac{m(2n+1)\pi}{N}\right]\right).
$$

(43)

Comparing the first term of (41) with first one of (43), it can be seen that $2\sum_{n=0}^{N-1} x(n)\left(\cos\left[\frac{m(2n+1)\pi}{N}\right]\right)$ is decimated and asymmetrically extended of (41) with index $m=0:N-1$. Similarly, $2\sum_{n=0}^{N-1} x(n)\left(\sin\left[\frac{m(2n+1)\pi}{N}\right]\right)$ is decimated and symmetrically extended of (41) with index $m=1:N$. It is observed that proper zero padding of the sequences, symmetric convolution can be used to perform linear convolution. The circular convolution of cosine and sine periodic sequences in time/spatial domain is equivalent to multiplication in the DFT domain. Then, the DFT matrix is resulted from the DCT-II and DST-II matrices.

3.4. Unified hybrid fast algorithm

Based on the above conversions from the proposed decomposition of DCT-II, we can form a hybrid fast algorithm that can cover DCT-II, DST-II, and DFT. The general block diagram of the proposed hybrid fast algorithm is shown in Fig. 4. The common recursive block of $[\mathbf{P}]_{N/2^h-1}\mathbf{L}\; blockdiagonal\,()[\mathbf{I}_2 \otimes \mathbf{Z}_2]Rblockdiagonal\,()\begin{bmatrix}\mathbf{I}_2 & \mathbf{I}_2\\ \mathbf{I}_2 & -\mathbf{I}_2\end{bmatrix}\mathbf{Q}_{N/2^h-1}$ is multiplied repeatedly according to the size of the kernel with different transforms as like as bracket $((((\cdot))))$. The requiring computational complexity of individual DCT-II, DST-II, and DFT is summarized in Table 1 and Table 2. It can be seen that the proposed hybrid algorithm requires little more computations in addition and multiplication compared to Wang's result [13]. However, the proposed scheme requires a much less computational complexity in addition and multiplication compared to those of the decompositions proposed by [11,13,18]. In addition, compared to these transforms, the proposed hybrid fast algorithm can be efficiently extensible to larger transform sizes due to its diagonal block-wise inverse operation of recursive structure. Moreover, the proposed hybrid structure is easily extended to cover different applications. For example, a base station wireless communication terminal delivers a compressed version of multimedia data via wireless communications network. Either DCT-II or DST-II can be used in compressing multimedia data since the proposed decomposition is based on block diagonalization it can significantly reduce its complexity due to simple structure[11,19, 22], for various multimedia sources. The DCT image coding can be easily implemented in the proposed hybrid structure as shown in Fig. 4(b). From (45), the DCT-II is obtained by taking a real part of multiplication result of $e^{-j\pi m/2N}$ with $F_N = \{W^{nm}\}$. If the DCT-II is multiplied by $\bar{I}_N C_N M_N$, then we get DST. If the DCT and DST are convolved in time and frequency domain and multiplied by $2e^{-j\pi m/N}$, the DFT matrix can be obtained. Thus, the proposed hybrid algorithm enables the terminal to adapt to its operational physical device and size.

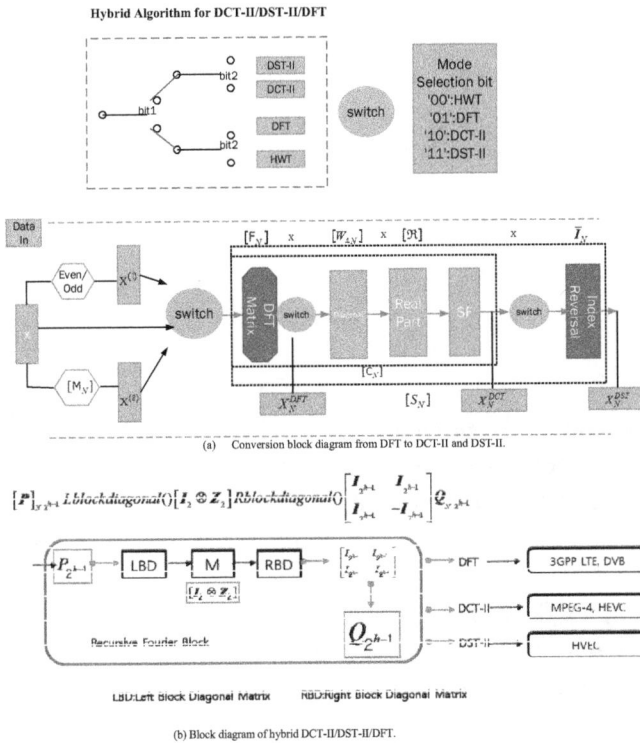

(a)   Conversion block diagram from DFT to DCT-II and DST-II.

(b) Block diagram of hybrid DCT-II/DST-II/DFT.

Figure 4. Recursive DCT-II/DST-II/DFT Structure Based on Jacket matrix.

3.5. Numerical simulations

As shown in [7] the coding performance DST outperforms DCT at high correlation values (ρ) and is very close to that of the KLT. Since the basis vectors of DCT maximize their energy distribution at both ends, hence the discontinuity appears at block boundaries due to quantization effects. However, since the basis vectors of DST minimizes their energy distribution at other ends, DST provides smooth transition between neighboring blocks. Therefore, the proposed hybrid transform coding scheme provides a consistent reconstruction and preserves more details, as shown in Fig. 6 with a size of 512 x 512 and 8 bits quantization.

Now consider an $N \times N$ block of pixels, X, containing $x_{i,j}$, i, $j=1, 2, \ldots, N$. We can write 2-D transformation for the kth block X as $Y_S = T_S Q X_k Q^T$ and $Y_C = T_C X_k$.

Depending on the availability of boundary values (in top- boundary and left-boundary) in images the hybrid coding scheme accomplishes the 2-D transform of a block pixels as two sequential 1-D transforms separately performed on rows and columns. Therefore the choice of 1-D transform for each direction is dependent on the corresponding prediction boundary condition.

- Vertical transform (for each column vector): employ DST if top boundary is used for prediction; otherwise use DCT.

- Horizontal transform (for each row vector): employ DST if left boundary is used for prediction; otherwise use DCT.

What we observed from numerical experiments is that the combined scheme over DCT-II only performs better in perceptual clarity as well as PSNR. Jointly optimized spatial prediction and block transform (see Fig. 5 (e) and (f)) using DCT/DST-II compression(PSNR 35.12dB) outperforms only DCT-II compression(PSNR 32.38dB). Less blocky artifacts are revealed compared to that of DCT-II. Without *a priori* knowledge of boundary condition, DCT-II performs better than any other block transform coding. The worst result is obtained using DST-II only.

4. Conclusion

In this book chapter, we have derived a unified fast hybrid recursive Fourier transform based on Jacket matrix. The proposed analysis have shown that DCT-II, DST-II, and DFT can be unified by using the diagonal sparse matrix based on the Jacket matrix and recursive structure with some characters changed from DCT-II to DST-II, and DFT. The proposed algorithm also uses the matrix product of recursively lower order diagonal sparse matrix and Hadamard matrix. The resulting signal flow graphs of DCT-II, DST-II, and DFT have a regular systematic butterfly structure. Therefore, the complexity of the proposed unified hybrid algorithm has been much less as its matrix size gets larger. This butterfly structure has grown by a recursive nature of the fast hybrid Jacket Hadamard matrix. Based on a systematic butterfly structure, a unified switching system can be devised. We have also applied the circulant channel matrix in our proposed method. Thus, the proposed hybrid scheme can be effectively applied to the heterogeneous transform systems having various matrix dimensions. Jointly optimized DCT and DST-II compression scheme have revealed a better performance (about 3dB) over the DCT or DST only compression method.

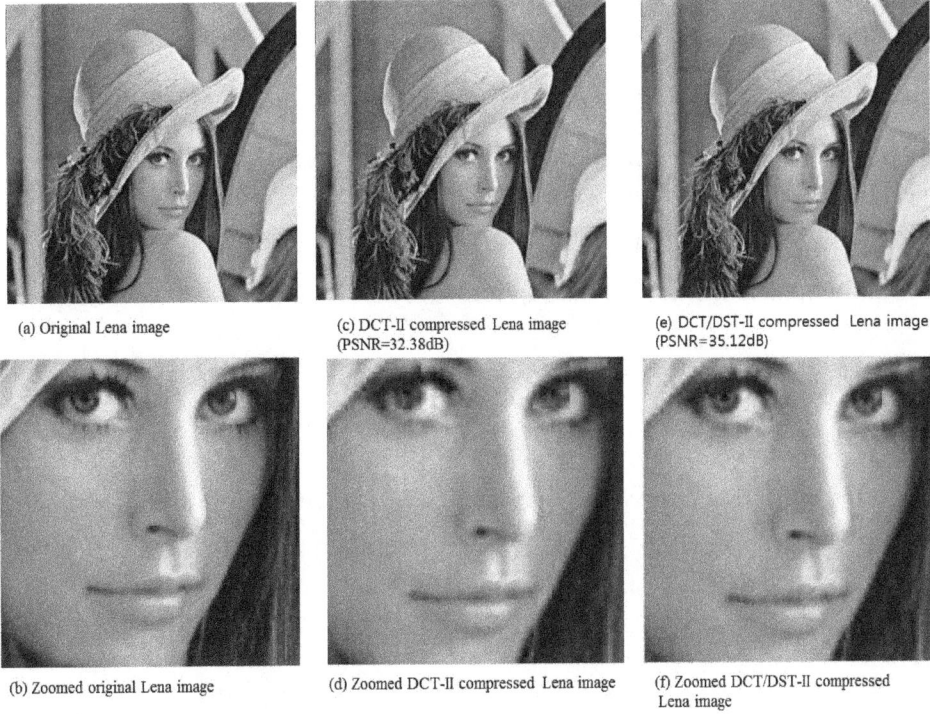

(a) Original Lena image

(c) DCT-II compressed Lena image (PSNR=32.38dB)

(e) DCT/DST-II compressed Lena image (PSNR=35.12dB)

(b) Zoomed original Lena image

(d) Zoomed DCT-II compressed Lena image

(f) Zoomed DCT/DST-II compressed Lena image

Figure 5. Image Coding Results showing DCT-II only and jointly optimized DCT/DST-II compression (a) Original Lena image (b) zoomed original Lena image (c) DCT-II compressed Lena image(PSNR=32.38 dB) (d) Zoomed DCT-II compressed Lena image (e) DCT/DST-II compressed Lena image (PSNR=35.12 dB) (f) Zoomed DCT/DST-II compressed Lena image.

Appendix

Appendix A

A Proof of Proposition 1

We use mathematical induction to prove Proposition 1. The lowest order BIJM is defined as

$$
J_8 = \begin{bmatrix} I_2 & I_2 & I_2 & I_2 \\ I_2 & -C_2 & C_2 & -I_2 \\ I_2 & C_2 & -C_2 & -I_2 \\ I_2 & -I_2 & -I_2 & I_2 \end{bmatrix} \tag{44}
$$

where $C_2 = \frac{H_2}{\sqrt{2}}$. Since

$$J_8^{-1} = \begin{bmatrix} I_2 & I_2 & I_2 & I_2 \\ I_2 & -C_2^T & C_2^T & -I_2 \\ I_2 & C_2^T & -C_2^T & -I_2 \\ I_2 & -I_2 & -I_2 & I_2 \end{bmatrix} \tag{45}$$

equation (4) holds for $2N = 8$. Now we assume that the BIJM J_N satisfies (4), i.e., $J_N J_N^T = \frac{N}{2} I_N$. Since $J_{2N} J_{2N}^T = (J_N \otimes H_2)(J_N \otimes H_2)^T = (J_N J_N^T) \otimes (H_2 H_2^T) = \frac{N}{2} I_N \otimes 2I_2 = N I_{2N}$, this proposition is proved by mathematical induction that (4) holds for all $2N$. If $N = 1$, certainly $J_2 J_2^T = I_2$.

Appendix B

A Proof of Lemma 1

According to the definition of an $N \times N$ matrix B_N, B_N is given as follows:

$$B_N = \begin{bmatrix} C_{4N}^{\Phi_0} & C_{4N}^{\Phi_1} & C_{4N}^{\Phi_2} & \cdots & C_{4N}^{\Phi_{N-1}} \\ C_{4N}^{(2k_0+1)\Phi_0} & C_{4N}^{(2k_0+1)\Phi_1} & C_{4N}^{(2k_0+1)\Phi_2} & \cdots & C_{4N}^{(2k_0+1)\Phi_{N-1}} \\ C_{4N}^{(2k_1+1)\Phi_0} & C_{4N}^{(2k_1+1)\Phi_1} & C_{4N}^{(2k_1+1)\Phi_2} & \cdots & C_{4N}^{(2k_1+1)\Phi_{N-1}} \\ \vdots & \vdots & \vdots & \cdots & \vdots \\ C_{4N}^{(2k_{N-2}+1)\Phi_0} & C_{4N}^{(2k_{N-2}+1)\Phi_1} & C_{4N}^{(2k_{N-2}+1)\Phi_2} & \cdots & C_{4N}^{(2k_{N-2}+1)\Phi_{N-1}} \end{bmatrix} \tag{46}$$

where $k_i = i + 1$. Since $\cos((2k+1)\Phi_m) = 2\cos(2k\Phi_m)\cos(\Phi_m) - \cos((2k-1)\Phi_m)$, we have

$$C_{4N}^{(2k_i+1)\Phi_m} = -C_{4N}^{(2k_i-1)\Phi_m} + 2C_{4N}^{(2k_i)\Phi_m} C_{4N}^{\Phi_m}. \tag{47}$$

Using (47), B_N can be decomposed as:

$$B_N = L_N \begin{bmatrix} \frac{1}{\sqrt{2}} & \frac{1}{\sqrt{2}} & \frac{1}{\sqrt{2}} & \cdots & \frac{1}{\sqrt{2}} \\ C_{4N}^{2k_0\Phi_0} & C_{4N}^{2k_0\Phi_1} & C_{4N}^{2k_0\Phi_2} & \cdots & C_{4N}^{2k_0\Phi_{N-1}} \\ C_{4N}^{2k_1\Phi_0} & C_{4N}^{2k_1\Phi_1} & C_{4N}^{2k_1\Phi_2} & \cdots & C_{4N}^{2k_1\Phi_{N-1}} \\ \vdots & \vdots & \vdots & \cdots & \vdots \\ C_{4N}^{2k_{N-2}\Phi_0} & C_{4N}^{2k_{N-2}\Phi_1} & C_{4N}^{2k_{N-2}\Phi_2} & \cdots & C_{4N}^{2k_{N-2}\Phi_{N-1}} \end{bmatrix} D_N \tag{48}$$

$$= L_N X_N D_N$$

which proves (10) in Lemma 1.

Appendix C

A Proof of Equation (18)

By using the sum and difference formulas for the sine function, we can have

$$
\begin{aligned}
S_{4N}^{(2k_{N-1})\Phi_j} &= C_{4N}^{\Phi_j} = S_{4N}^{\Phi_{N-j-1}}, S_{4N}^{(2k_i-1)\Phi_j} = 2S_{4N}^{2k_i\Phi_j}C_{4N}^{\Phi_j} - S_{4N}^{(2k_{i+1}-1)\Phi_j}, \\
S_{4N}^{(2k_{i+1}-1)\Phi_j} &= 2S_{4N}^{2k_i\Phi_j}C_{4N}^{\Phi_j} - S_{4N}^{(2k_i-1)\Phi_j}, \\
S_{4N}^{(2k_0-1)\Phi_j} &= \left(2S_{4N}^{(2k_0-1)\Phi_j} - 2S_{4N}^{(2k_1)\Phi_j} + \cdots + 2S_{4N}^{(2k_{N-2}-1)\Phi_j}\right)C_{4N}^{\Phi_j} \\
&= 2S_{4N}^{(2k_0-1)\Phi_j}C_{4N}^{\Phi_j} - S_{4N}^{(2k_1-1)\Phi_j}, \\
S_{4N}^{(2k_1-1)\Phi_j} &= \left(2S_{4N}^{(2k_1-1)\Phi_j} - 2S_{4N}^{(2k_1)\Phi_j} + \cdots + 2S_{4N}^{(2k_{N-2}+1)\Phi_j}\right)C_{4N}^{\Phi_j} \\
&= 2S_{4N}^{(2k_1-1)\Phi_j}C_{4N}^{\Phi_j} - S_{4N}^{(2k_2-1)\Phi_j}, \\
&\vdots \\
S_{4N}^{(2k_{N-3}-1)\Phi_j} &= \left(2S_{4N}^{(2k_{N-3}-1)\Phi_j} - 2S_{4N}^{(2k_{N-2})\Phi_j} + 1\right)C_{4N}^{\Phi_j} \\
&= 2S_{4N}^{(2k_{N-3}-1)\Phi_j}C_{4N}^{\Phi_j} - S_{4N}^{(2k_{N-2}-1)\Phi_j}, \\
S_{4N}^{(2k_{N-2}-1)\Phi_j} &= \left(2S_{4N}^{(2k_{N-2})\Phi_j} - 1\right)C_{4N}^{\Phi_j} \\
&= 2S_{4N}^{(2k_{N-2})\Phi_j}C_{4N}^{\Phi_j} - S_{4N}^{(2k_{N-1}-1)\Phi_j}
\end{aligned}
\tag{49}
$$

where $k_i = i+1$, $\Phi_j = 2j+1$, i, $j = 0, 1, \cdots, N-1$.

By taking (49) and into the right hand side of (18), we have

$$
U_N Y_N D_N =
\begin{bmatrix}
S_{4N}^{(2k_0-1)\Phi_0} & S_{4N}^{(2k_0-1)\Phi_1} & \cdots & S_{4N}^{(2k_0-1)\Phi_{N-1}} \\
S_{4N}^{(2k_1-1)\Phi_0} & S_{4N}^{(2k_1-1)\Phi_1} & \cdots & S_{4N}^{(2k_1-1)\Phi_{N-1}} \\
\vdots & \vdots & \ddots & \vdots \\
S_{4N}^{(2k_{N-1}-1)\Phi_0} & S_{4N}^{(2k_{N-1}-1)\Phi_1} & \cdots & S_{4N}^{(2k_{N-1}-1)\Phi_{N-1}}
\end{bmatrix}.
\tag{50}
$$

The left hand side of (18) matrix $[A]_N$ from $[Y]_N$ can be represented by

$$
A_N =
\begin{bmatrix}
S_{4N}^{(2k_0-1)\phi_0} & S_{4N}^{(2k_0-1)\phi_1} & \cdots & S_{4N}^{(2k_0-1)\phi_{N-1}} \\
S_{4N}^{(2k_1-1)\phi_0} & S_{4N}^{(2k_1-1)\phi_1} & \cdots & S_{4N}^{(2k_1-1)\phi_{N-1}} \\
\vdots & \vdots & \ddots & \vdots \\
S_{4N}^{(2k_{N-1}-1)\phi_0} & S_{4N}^{(2k_{N-1}-1)\phi_1} & \cdots & S_{4N}^{(2k_{N-1}-1)\phi_{N-1}}
\end{bmatrix}.
\tag{51}
$$

We can obtain (50) and (51) are the same and the expression of (18) is correct.

Acknowledgements

This work was supported by MEST 2012- 002521, NRF, Korea.

Author details

Daechul Park[1*] and Moon Ho Lee[2]

*Address all correspondence to: fia4joy@gmail.com

1 Hannam University, Department of Information and Communication engineering, Korea

2 Chonbuk National University, Division of Electronics and Information Engineering, Korea

References

[1] Rao, KR. and Yip, P., *Discrete Cosine Transform: Algorithms, Advantages, Applications*. Boston, MA: Academic Press, 1990.

[2] Richardson, IE., *The H.264 Advanced Video Compression Standard*, 2nd ed. Hoboken, New Jersey: John Wiley and Sons.

[3] Rao, KR., Kim, DN., and Hwang, J. J., *Fast Fourier Transform: Algorithm and Applications*. New York, N.Y.: Springer, 2010.

[4] Jain, A. K., *Fundamentals of Digital Image Processing*. Prentice Hall, 1987.

[5] Wang, R., *Introduction to Orthogonal Transforms: With Applications in Data Processing and Analysis*. Cambridge, UK: Cambridge University Press, 2012.

[6] ITU-T SG16 WP3/JCT-VC, CE 7.5, "Performance analysis of adaptive DCT/DST selection," July 2011.

[7] Hai, J., Saxena, A., Melkote, V., and Rose, K., "Jointly optimized spatial prediction and block transform for video and image coding," *IEEE Trans. Image Process.*, vol. 21, no. 4, pp. 1874–1884, April 2012.

[8] Strang, G. and Nguyen, T., *Wavelets and Filer Banks*. Wellesley, MA: Wellesley-Cambridge Press, 1996.

[9] Lee, MH., "A new reverse Jacket transform and its fast algorithm," *IEEE Trans. Circuits Syst. II*, vol. 47, no. 1, pp. 39–47, Jan. 2000.

[10] Chen, Z.,. Lee, MH, and Zeng, G., "Fast cocyclic Jacket transform," *IEEE Trans. Signal Process.*, vol. 56, pp. 2143–2148, May 2008.

[11] Lee, MH., *Jacket Matrices-Construction and Its Application for Fast Cooperative Wireless Signal Processing*. LAP LAMBERT Academic publishing, Germany, November, 2012.

[12] Wang, CL. and Chen, CY., "High-throughput VLSI architectures for the 1-D and 2-D discrete cosine transform," *IEEE Trans. Circuits Syst. Video Technol.*, vol. 5, pp. 31–40, Feb. 1995.

[13] Wang, Z., "Fast Algorithm for the Discrete W Transform and for the Discrete Fourier Transform," *IEEE Trans. on Acoustics, Speech and Signal Process.*, vol. 32, No. 4, pp. 803 – 816, Aug. 1984.

[14] Lee, MH., "High speed multidimensional systolic arrays for discrete Fourier transform," *IEEE Trans. Circuits Syst. II*, vol. 39, no. 12, pp. 876–879, Dec. 1992.

[15] Kim, KJ., Fan, Y.,Iltis, R. A., Poor, H. V., and Lee, M. H., "A reduced feedback precoder for MIMO-OFDM cooperative diversity system," *IEEE Trans. Veh. Technol.*, vol. 61, pp. 584–596, Feb. 2012.

[16] Jang, U.,Cho, K.,Ryu, W., and Lee, HJ., "Interference management with block diagonalization for macro/femto coexisting networks," *ETRI Journal*, vol. 34, pp. 297–307, June 2012.

[17] Hou, HS., "A fast recursive algorithm for computing the discrete cosine transform," *IEEE Trans. Acoust., Speech, Signal Process.*, vol. 35, no. 10, pp. 1455–1461, Oct. 1987.

[18] Chen, WH., Smith, CH., and Fralick, SC., "A fast computational algorithm for the discrete cosine transform," *IEEE Trans. Commun.*, vol. 25, no. 9, pp. 1004–1009, Sep. 1977.

[19] Spencer, Q. H., Lee, A. Swindlehurst, M. Haardt, "Zero-forcing methods for downlink spatial multiplexing in multiuser MIMO channels", *IEEE Trans. Signal Process.*, vol. 52,no. 2, Feb. 2004.

[20] Andrews, HC., Caspari, KL., "A generalized technique for spectral analysis," *IEEE Trans. Computers*, vol. 19, no. 1, pp.16-17, 1970.

[21] Reju, VG.,Koh, SN.,Soon, IY., "Convolution using discrete sine and cosine transforms", IEEE Signal Processing Letters, vol. 14, no. 7, July 2007.

[22] Lee, MH., Khan, MHA., Sarker, MA. L., Guo, Y. and Kim, KJ., "A MIMO LTE precoding based on fast diagonal weighted Jacket matrices", *Fiber and Integrated Optics, Taylor and Francis*, Invited paper, vol. 31, no. 2, pp. 111-132, March 2012.

[23] Khan, MHA., Li, J., Lee, MH., "A block diagonal Jacket matrices for MIMO broadcast channel" *IEEE International Symposium on Broadband Multimedia Systems and Broadcasting*, Brunel University, June 4-7th, 2013, UK.

Permissions

The contributors of this book come from diverse backgrounds, making this book a truly international effort. This book will bring forth new frontiers with its revolutionizing research information and detailed analysis of the nascent developments around the world.

We would like to thank all the contributing authors for lending their expertise to make the book truly unique. They have played a crucial role in the development of this book. Without their invaluable contributions this book wouldn't have been possible. They have made vital efforts to compile up to date information on the varied aspects of this subject to make this book a valuable addition to the collection of many professionals and students.

This book was conceptualized with the vision of imparting up-to-date information and advanced data in this field. To ensure the same, a matchless editorial board was set up. Every individual on the board went through rigorous rounds of assessment to prove their worth. After which they invested a large part of their time researching and compiling the most relevant data for our readers.

The editorial board has been involved in producing this book since its inception. They have spent rigorous hours researching and exploring the diverse topics which have resulted in the successful publishing of this book. They have passed on their knowledge of decades through this book. To expedite this challenging task, the publisher supported the team at every step. A small team of assistant editors was also appointed to further simplify the editing procedure and attain best results for the readers.

Apart from the editorial board, the designing team has also invested a significant amount of their time in understanding the subject and creating the most relevant covers. They scrutinized every image to scout for the most suitable representation of the subject and create an appropriate cover for the book.

The publishing team has been an ardent support to the editorial, designing and production team. Their endless efforts to recruit the best for this project, has resulted in the accomplishment of this book. They are a veteran in the field of academics and their pool of knowledge is as vast as their experience in printing. Their expertise and guidance has proved useful at every step. Their uncompromising quality standards have made this book an exceptional effort. Their encouragement from time to time has been an inspiration for everyone.

The publisher and the editorial board hope that this book will prove to be a valuable piece of knowledge for researchers, students, practitioners and scholars across the globe.

List of Contributors

Francisco J. Mendoza-Torres and Juan Alberto Escamilla-Reyna
Facultad de Ciencias Físico-Matemáticas, Benemérita Universidad Autónoma de Puebla, Puebla, Mexico

Ma. Guadalupe Morales-Macías
Departamento de Matemáticas, Universidad Autónoma Metropolitana-Iztapalapa, México, D. F., Mexico

Salvador Sánchez-Perales
Instituto de Física y Matemáticas, Universidad Tecnológica de la Mixteca, Oaxaca, Mexico

Takashi Gyoshin Nitta
Department of Mathematics, Faculty of Education, Faculty of Education, Mie University, Japan

Nafya Hameed Mohammad
Department of Mathematics, College of Education, Salahaddin University, Erbil, Iraq

Massoud Amini
Department of Mathematics, Faculty of Mathematical Sciences, Tarbiat Modares University, Tehran, Iran

Adrian Bot and Nicolae Aldea
National Institute for Research and Development of Isotopic and Molecular Technologies, Cluj-Napoca, Romania

Florica Matei
University of Agricultural Sciences and Veterinary Medicine, Cluj-Napoca, Romania

Renbiao Wu, Qiongqiong Jia and Wenyi Wang
Tianjin Key Lab for Advanced Signal Processing, Civil Aviation University of China, Tianjin, P. R. China

Jie Li
Department of Electronic and Information Engineering, Zhonghuan Information College Tianjin University of Technology, Tianjin, P. R. China

Andrej Prošek and Matjaž Leskovar
Reactor Engineering Division, Jožef Stefan Institute, Ljubljana, Slovenia

Daechul Park
Hannam University, Department of Information and Communication engineering, Korea

Moon Ho Lee
Chonbuk National University, Division of Electronics and Information Engineering, Korea

Index

www.ingramcontent.com/pod-product-compliance
Lightning Source LLC
Chambersburg PA
CBHW062004190326
41458CB00009B/2962